THE
UNSELFISH GENOME

HOW

DARWIN & DAWKINS

DID NOT SEE

<u>THE MISSING HALF</u>

OF

THE THEORY OF EVOLUTION

&

NEW RESEARCH REVEALS THE HORMONES

THAT CONTROL HUMAN AGING

Jeff T. Bowles

PREFACE: 4

INTRODUCTION 8

CHAPTER 1: THE FIRST GLIMPSE-EXCEPTIONALLY LONG-LIVED ANIMALS 29

CHAPTER 2: THE SECOND PIECE OF THE PUZZLE –RAPIDLY AGING ANIMALS. 57

CHAPTER 3: THE THIRD PUZZLE PIECE-LIFETIME HUMAN REPRODUCTIVE HORMONE CHANGES 72

CHAPTER 4: THE FOURTH PUZZLE PIECE- ASEXUAL VERSUS SEXUAL REPRODUCTION 92

CHAPTER 5: THE FIFTH PUZZLE PIECE- MALE & FEMALE SEX TYPES 120

CHAPTER 6: THE SIXTH PUZZLE PIECE-HOMOSEXUALITY IN 152
ANIMALS & HUMANS 152

CHAPTER 7: THE 7TH PUZZLE PIECE- HUMAN MENOPAUSE & AGE-CHANGES IN FERTILITY IN VARIOUS ANIMALS 179

CHAPTER 8: PUZZLE PIECE #8-THE EXISTENCE OF AGING GENES 209

CHAPTER 9: PUZZLE PIECE #9 AGING CAN BE SLOWED & EVEN REVERSED. 232

CHAPTER 10: THE TIMING OF THE EXPRESSION OF AGING GENES-VIA 5MC, CHROMATIN, LAMIN A, WRN 250

CHAPTER 11: HOW SEX AND AGING ARE SELECTED FOR- EVERYWHERE & ALL THE TIME 275

CHAPTER12: A WORLD WITHOUT MEN-THE FUTURE COURSE OF EVOLUTION? 301

BONUS CHAPTER #13- THE EVOLUTION OF HUMAN RELIGION & RELIGIOUS SUICIDE 305

BONUS CHAPTER #14 WHAT CHEMICALS AND MOLECULES THAT ARE GOOD FOR YOU HAVE IN COMMON 323

APPENDIX A: MY THOUGHTS ON CHARLES DARWIN & SOME DELETED TEXT 345

APPENDIX B: AGEING: THEORY NEEDS TO BE REVISED 350

APPENDIX C: GUIDE: SEVEN INCREDIBLY OLD MOJAVE DESERT PLANTS 355

APPENDIX D: AGING GENE STUDY..362

APPENDIX E: ALL 11 ABSTRACTS MENTIONING SPECIES LEVEL SELECTION IN THE ENTIRE PUB MED SCIENCE DATABASE AS OF 12/2015. 364

APPENDIX F: IT TURNS OUT WERNER'S PROTEIN (WRN) SHUTS OFF AGING GENES JUST LIKE ALL THE OTHER ANTI-AGING PROTEINS. 372

APPENDIX G: THE CONSERVATIVE US GOVERNMENT'S NIH IS NOW PROMOTING THE LH/ALZHEIMER'S DISEASE CONNECTION!........................375

Preface:

At the end of this book you will not only be all the wiser from the perspective of evolutionary theory; you will see where Darwin and Dawkins were wrong because of the large portion of evolutionary theory that they did not see. Yes Darwin, and his theoretical heir, Dawkins, were both half right, but to me, being only half right is the same as being wrong. Darwin, a true hero, will remain blameless and remain a Demi-God of biology after the publication of this book, as he did not have the information necessary to see the other half of the theory of evolution.

Dawkins and his contemporaries, however, have had the facts required to complete the theory of evolution, but have chosen to ignore them.

While reading this book, you will likely experience quite a few…"AHA!.." moments, and when you see the final direction of where human and all other life forms' evolution is headed you will be likely be quite shocked-it really is a strange new world that has been uncovered!

You will also see that correctly completing the theory of evolution is not just an academic exercise. It has tangible predictions about how to manage your hormones to slow, and for some periods stop, and at times even reverse the aging process. Finally some evolutionary theory with practical applications!

While the topic of this book seems like it should be presented in a scholarly, scientific manner for the benefit of any evolutionary biologists who will be evaluating it, I have learned over my lifetime of interaction with biologists and scientists that this book would fall mostly on deaf ears. Unlike the days of Darwin, science today is mostly conducted by highly paid professors with a large investment in being right and a huge risk if they are wrong. Today's professional biologists will be the last to embrace this book, just you wait and see. So rather than make it a dry, scientific droning monologue, this book is written to be fun, and entertaining, and it has lots of wonderful pictures. Parts of the book may seem politically incorrect, and I guess they are. Sorry, but evolution , like the truth, is politically incorrect at times.

And yes, I know that I misspelled the word genome on the cover of the book. Instead I spelled it geneome with an extra e.

(Actually in this paperback version, the Amazon editors forced me to spell it correctly-so the following only applies to the ebook version).

This was not an accident, but instead allowed me to retain the idea of the selfish gene which does drive much of evolution, inside a broader category of the unselfish genome, which explains the rest of evolution. Thus by misspelling the word genome, the title of the book says it all…

THE UNSELFISH GENEOME

where the smaller concept of selfish gene is contained within the bigger concept of the unselfish geneome. In fact, when they invented the word genome, they may have made a mistake; I think spelling it geneome would be better, it is more correct as it contains the entire word gene in it rather than just "gen". I hereby place my vote to ask Webster's Dictionary to change the official spelling of genome to geneome sometime in the future.

Also on the cover I single out Charles Darwin and Richard Dawkins as the theorists that I am trying to correct with this completion of the theory of evolution. Charles Darwin stands by himself, and other than his being only half right, I do not say he was wrong in any way, but rather, his version of the theory was just incomplete. The idea of the gene (discrete packets of inheritable information) was not known by science until 20 years after Darwin died in 1882. It was around 1900 when the forgotten work of Gregor Mendel, who published his concept of the gene in pea plants in 1865, was rediscovered by science.

On the other hand, when I target Richard Dawkins in the title, I am not just targeting him. Dawkins represents all the theorists who came after Darwin, who added to Darwin's theory from the perspective of the primacy of the selfish gene. By focusing on just the gene trying to make more copies of itself, these new evolutionists were able to apply all sorts of mathematical acrobatics towards trying to explain evolution. They tried to turn the "soft" science of biology and evolution into a "hard" science of the mathematically describable selfish gene. So when I call out Dawkins to the mat, I am also calling out the entire mainstream selfish gene theory community from Medawar and Williams, to Fisher, Wright, Smith, Haldane, Hamilton, and Charlesworth, etc.

Dawkins was just the most famous popularizer of all their work for the masses. He represents them all for my purposes, and that is why I am saying Dawkins is not only incomplete, but actually quite wrong on many points. It turns out, that Dawkins will be right sometime far in the evolutionary future.

You might say he is millions if not billions of years ahead of his time! Until then, there are a few stumbling blocks in front of his version of evolutionary theory that need to be overcome. They are overcome in this book. With that out of the way let's get on to the argument. I will begin as Dawkins did in his book The Selfish Gene, using the musings of a child to begin his dialogue.

Introduction

When I was a kid, my family had a dog named Otto. He was a German short-haired pointer- a medium sized dog as far as dogs go. We got him when I was 7. He looked like this.

By the time I was 16, not even finished with puberty, Otto was getting old. His hair was graying around his chin, and he was starting to get arthritis, He was slower and less energetic than in his younger years. By the time I was 19, he had to be put down because he had grown so decrepit that he was suffering more than enjoying life. He was about 12 years old in human years, but an old man of 84 in dog years-while I was still not even an adult. I still could not legally even buy myself a beer.

This truly impressed upon me, from a young age, how aging just had to be programmed. How else could two animals, such as me a human, and Otto a dog, being made of almost the identical flesh and blood, age at such dramatically different rates!? From this simple common sense example, I always assumed that aging was programmed and that everyone in the science world must have known it as well.

For reasons I will explain in a later chapter, when I was around the age of 28, I decided to change my life's direction and to spend most of my time, rather than working for money, trying to figure out how to find a cure for aging.

Almost 30 years old, I re-entered college as a freshman in biology, taking my first class, Bio 101, looking to study anything related to the <u>facts</u> about aging. For some reason I just assumed that the theory part of aging had already been settled and that it was definitely programmed-thus I had no interest in theory, rather, I was only interested in what I could divine from the facts.

I was quite gung-ho to spend as much of my time studying aging as possible, but I was limited, because I did not know anything about genes , genetics, or biochemistry. I initially spent much more of my time learning about all the facts of aging that did not require a mastery of the harder biological sciences. I LEARNED ABOUT EVOLUTION FROM THE TOP DOWN- I started with the BIG PICTURE, and learned the small picture stuff like biochemistry, genes, and DNA last.

While waiting to get through the prerequisite courses over the years that would let me study genetics and organic chemistry, I was "stuck", forced to study the BIG PICTURE stuff. But I did it with an obsessive zeal.

I started learning about the higher level facts about aging, like

-the differences of life span between species,

-examples of organisms with dramatically long life spans,

-examples of organisms that undergo rapid aging and death at the end of their lives

-examples of animals / organisms that do not seem to get physically older no matter how long they live

-examples of animals, as complicated as turkeys, that can clone themselves (females only) without the need for sexual reproduction

-how hormones change with age,

-various regimens that can slow the aging process like semi-starvation (also known as caloric restriction)

-substances, supplements, hormones, surgical procedures that had increased life span in some species

With my naïve and ignorant assumption that aging was programmed, in 1998 I was able to make various predictions that were considered laughable on the day of publication of my first science paper, but have been coming true with a vengeance since then.

One outrageous prediction at the time from my 1998 paper was that a hormone involved in human reproduction, luteinizing hormone, would be found to be involved in causing aging and Alzheimer's in particular. At the time of the prediction, science believed that receptors for luteinizing hormone were only to found in sex-related tissues of humans. What a shock to the skeptics a few years later when the receptors were found all over the body and throughout the brain!

This is no longer some wild prediction as scientists at the US government's National Institute of Health (NIH) a few years ago also got on the LH/Alzheimer's bandwagon by conceding the link between LH and Alzheimer's:

Gonadotropin-releasing hormone receptor system: modulatory role in aging and neurodegeneration. Wang L, Chadwick W, Park SS, Zhou Y, Silver N, Martin B, Maudsley S. *CNS Neurol Disord Drug Targets.* 2010 Nov;9(5):651-60 (see the full abstract in the appendices).

Another crazy prediction made in the 1998 paper was that a little thing called epigenetics (which is just a fancy name for things like proteins or molecules that cover up your genes and prevent them from being read-like insulation covering a wire) would be found to be the major controlling factor that drove the aging program. It only took 12 years for the first papers by others in the aging field to start making the same prediction and providing the evidence, see-

Exp Gerontol. 2010 Apr; 45(4): 253-254.

Epigenetics and Aging Thomas A. Rando

Shortly after my first paper hit the press in 1998, I quickly found out that <u>my approach to studying aging was completely backwards</u> compared to all the other researchers who were studying and writing about aging! I was informed by the most eminent aging scientists in the world that I didn't know the first thing about aging, and that it was settled long ago, that aging is NOT programmed , but just an accidental artifact of evolution. Knowing I was right, deep in my bones, I was having none of this. That was when I went back to their old theory papers and started reading them with a purpose-to figure out and show them where they were wrong.

Almost all career biologists work their way up, beginning with the ultra-complicated low level facts about how DNA works, via genes, meiosis, mitosis, recombination, chromosomes, crossovers, etc., etc. Only when they become PhD candidates, or professors do most mainstream aging theorists start dealing with the BIG PICTURE items.

So in the normal course of an aging theorist's career, he or she begins with a complete brainwashing at the genetic level with the mainstream view, and only then does he or she look at the BIG PICTURE. At this point in their careers they "know for a fact" that aging is not programmed, and that only things that are good for you which help you spread your genes can evolve and be selected for. They know that is impossible for a gene that evolved to <u>intentionally</u> cause an individual to grow old and die to exist!

Amazingly, I did not find out that accidental aging was the mainstream view until after I had published my first paper on aging in 1998 whose basic premise was that aging was programmed. Amazingly, up until the point I did not even realize that there was any controversy about aging being programmed vs non-programmed! I had studied all the facts and created a unified theory of aging that also necessarily implied a new view of how evolution works completely in a vacuum! I was even accused of this by none other than the famous biologist Robin Holliday after he read my manuscript. I did not realize how far off base I was until I was getting very negative feedback about how ridiculous my theory was by every mainstream aging theorist who read it!

In retrospect, I was quite naïve and ridiculously overconfident to have never read any of the theory from the mainstream biologists about evolution, genetics, and aging. I did not read Darwin, I had not read Medawar nor Williams (whose early papers comprise the two major foundations of modern aging theory) nor had I read the math jocks-Hamilton and Fisher, nor the modern Kirkwood and Charlesworth (who correctly told me that I didn't know how (his idea of) evolution worked!) nor had I even read the book of the popularizer Dawkins and his Selfish Gene book.

But actually it was quite fortunate that I hadn't. Because if I did, I might have been persuaded that aging was, as everyone else believed, non-programmed-just an accident of an animal or organism living longer than it was designed for in the wild. Had I ever been persuaded of this, I never would have completed my unified theory of aging paper, and this book you are now reading would never have been written by me. In my case ignorance truly was bliss! And now I will show you in this book that Stephen Hawking had it right when he coined the phrase " The greatest enemy of knowledge is not ignorance, it is the illusion of knowledge." I will make the case that almost all mainstream evolutionary and aging theorists suffer from the illusion of knowledge. Shortly I will start by challenging excerpts from Richard Dawkins' bestselling book "The Selfish Gene".

Now let me walk what I said back just a little bit. I am not saying that today's mainstream theorists in evolution and aging are wrong. Actually they are all half right. Their only problem is they are missing the second half of evolution that you need to know before you can understand everything and explain all the many exceptions that exist in their half-right theories. I predict that someday in the far future, after evolution has worked out some of its oddities, maybe in a billion years the selfish gene theorists will finally be correct, but until then, they are only half right-you will see what I mean soon enough.

So at age 28 I began a quest that has not stopped to this day to acquire all the facts I can about aging and with them, solve the puzzle and discover how and why we age. This quest went into high gear from age 28 to 38 where I spent most of my time either in various college biology, chemistry, and genetics classes or mostly in the medical school library reading every science article and study related to aging and aging-related diseases.

For example, if I found out that the incidence of cancer increases dramatically with old age, I might spend a year or two studying cancer with the only thought in mind as to how it related to aging.

When I discovered that some animals lived exceptionally long lives based on their body size, I might have spent another year trying to see what all these animals had in common.

Upon hearing of animals and organisms that rapidly age and die right after reproducing, I would study them in detail.

If I encountered diseases in humans like progeria or Werner's syndrome where aging was much faster than normal, I would spend another year just studying those diseases in excruciating detail.

If I ever encountered any talk about how aging was not programmed, that it was due to just accidental wear and tear, I guess I just ignored it, I knew in my bones that aging was programmed! One way of looking at it is that ignoring all prior theories about aging might have been considered justified by the fact that almost all theories had been constructed many years in the past when there were very few facts to work with compared to the modern era in which I was theorizing.

So I ended up doing for years what I used to like to do as a child, doing a huge jigsaw puzzle on the dining room table- but in this case it was a puzzle at the Northwestern medical school library computer cubicle. The people who worked there rightfully must have thought I was crazy…and maybe I was…but being a Northwestern Alumni (from business school) I had an alumni card…..and all they really could do was snicker behind my back (am I hearing voices??).

By the time I had my Eureka moment at the genetic level of the puzzle I had been obsessively working on the puzzle for maybe 8 years. There were two things that came together at the same time that pulled the blindfold off my eyes.

The first breakthrough came when I was reading about how cells looked at the microscopic level from the children with the rapid aging disease called Progeria where they die of old age by their 12th year on average.

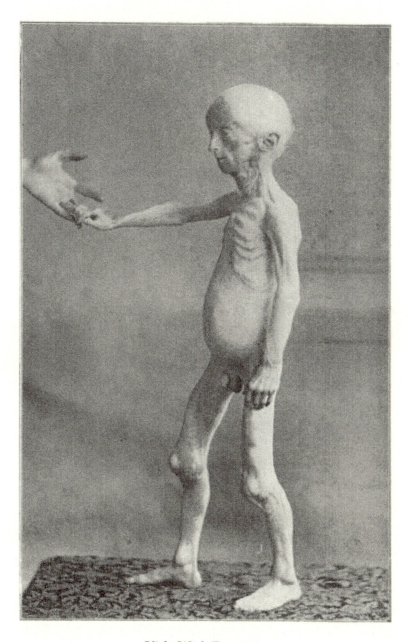

Kids With Progeria

Kids With Progeria

The article discussed how progeria cells had DNA that was very disorganized and almost haphazardly stuffed inside the cell nucleus instead of being tightly wound up in neat little bundles and all stuffed into the cell with extreme precision. Progerics' DNA was described almost like it was a terribly tangled unwashed fright wig as opposed to neatly cropped and beautifully organized headful of beautiful braids. Also their cells had misshapen nuclei:

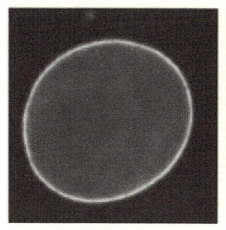 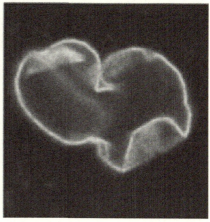

Normal Nucleus Vs Progeria Nucleus

And the second breakthrough at the genetic level of my understanding was that another rapid aging disease that kicks in at puberty called Werner's syndrome contained all the symptoms of rapid aging as seen in Progeria plus a whole menu of other rapid aging symptoms (seen in normally aging humans) unique to Werner's syndrome alone (more on this later).

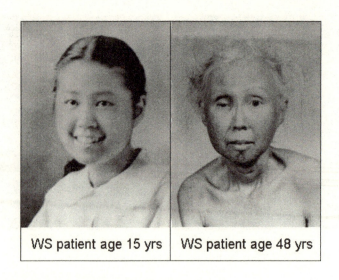

WS patient age 15 yrs | WS patient age 48 yrs

What I figured out from these two observations is what I had been starting to suspect-that much of **aging is caused by the loss of substances that attach to your DNA** (proteins, molecules, and other things) to prevent aging genes (designed to kill you) from being activated (which has turned out to be true), AND that aging is not a single system that ages you but is composed of multiple aging systems that age you independently of each other (or the concept of multiple aging systems).

So the two diseases of Progeria and Werner's syndrome were the inspiration for my completion of the unified theory of aging at the genetic and molecular level. However, prior to that, I had several major insights at the BIG PICTURE level which shed light on how the smaller puzzle of genetics fit into the BIG PICTURE of the evolutionary purpose of aging.

One major insight I had was that from a broader perspective, gained by studying the aging rates of various animals, was that a certain class of animals tended to have exceptionally long lifespans for their body size . These were plants and animals that either flew (bats and birds), were highly intelligent (humans and chimpanzees and apes), that had full body armor (like tortoises, lobsters, and clams), that were isolated from others (like deep dwelling fish, cave animals, animals that lived completely underground, or plants that lived on mountain tops or in the desert) and plants that contained defensive poisons throughout their biomass.

Another insight was that rapidly aging animals that might age and die within several days after living for years, all had one thing in common. They aged and died rapidly while in the process of reproducing. While my new view of aging and evolution allows for these rapidly aging organisms to be considered normal examples, mainstream theorists are so perplexed by these cases that they have invented a special category for them called semelparous aging that has nothing to do with other types of aging! From my new point of view of evolution, these rapidly aging organisms are not an exception to the correct theory of aging or evolution , but an excellent starting point for understanding all other cases of aging.

With these new insights in mind, I wrote a paper in 1996 that suggested I had come up with a working unified theory of aging that combined all the various theories of aging into a single theory and started sending it out to various science journals for publication. My first paper on aging was rejected by all the prestigious journals such as Lancet, Science, Nature, and even Experimental Gerontology.

And still at this point, luckily for me, I had never read a theory paper on THE EVOLUTION OF aging other than my own! I had never read "The Selfish Gene" by Richard Dawkins. Had I done so, and accepted mainstream theory as authoritative, I never would have discovered anything new about aging. And I definitely would not have been able to correct and complete Darwin's theory of evolution which just so happened to be solved simultaneously with solving the riddle of aging. So enough of this prelude…let's get down to the nitty gritty and complete and correct Darwin's theory of evolution.

I must admit that I have never read Richard Dawkin's famous book, The Selfish Gene, before. But now that I am writing a book that is titled to take advantage of the popularity of his classic, I have now begun reading it.

But before digging in and starting at page 1, I decided to look at the areas that I know Dawkins would not be able to explain with his bottom up theory derived from starting with the gene to explain all higher orders of biology and evolution. I simply went to the index and looked up aging, sex, and menopause.

Starting with aging, I see there are only locations in the book at page 333, at pages 40-42 and supposedly at page 135 (I read all around page 135 and aging was never mentioned). So let's see what was said about aging, sex and menopause. I know it will be a huge problem for his theory.

"The question of why we die of old age is a complex one, and the details are beyond the scope of this book" He then quickly examines a few competing theories and actually does make a prophetic observation ..

"For instance, suppose a substance S is more concentrated in the bodies of older individuals than of young individuals." And

"There might also be a substance "Y" a "label" for youth in the sense that it is more concentrated in young bodies than in old ones."

But at the end of his brief analysis of aging he just PFTA (plucks from thin air)-

"What matters for present purposes is that the gene-selection view of evolution has no difficulty in accounting for the tendency of individuals to die when they get old." And that's it...aging is brushed under the carpet and never mentioned again! Out of sight out of mind!

And then as quickly as aging is swept under the carpet...he then admits the existence of sexual reproduction and crossing over is "more difficult to justify". (Crossing over means the mixing up of the genes between the two parental chromosomes before making new sex cells like eggs and sperm. So any sperm or egg will have a random mix of genes from both the mother and father rather than being exclusively maternal or paternal genes). Evolution basically puts your genes (one set form your mother and one set from your father in a blender before letting you make sperm or egg cells from them.)

Dawkins goes on to say "Why did sex, that bizarre perversion of straightforward replication (of the genes), ever arise in the first place?" "What is the good of sex?"

"This is an extremely difficult question for the evolutionist to answer. Most serious attempts to answer it involve sophisticated mathematical reasoning. I am frankly going to evade it"

He does make a feeble attempt to explain the existence of sex as being caused by one gene, the sex gene, being more selfish than all the rest and forcing all the other genes to only have a 50% chance of being passed on from sex so that it can survive. What a preposterous contortion of his logic! He should have just stopped at "I am frankly going to evade it".

And finally, I know he will have a big problem explaining another thing that really restricts the spreads of the selfish gene-menopause! Let's see what he has to say about that. Checking the index menopause we get pages 126-127 and 2, 4.

Let's take a look: pages 1 through 8 make no mention of menopause.

Now 126-127:"there is something genetically deliberate about the menopause-that it is an adaptation. It is rather difficult to explain. He then goes on to promote the idea of the grandmother hypothesis where menopause supposedly is selected for to increase the woman's overall spread of her genes by assisting the raising of her grandchildren and increasing their odds of survival who have 1/4th her genes as opposed to raising her own children that still that share 1/2 of her genes. This ridiculous idea just stinks of contrivance and one study has shown it to be false, see-

Hill, K.; Hurtado, A.M. (1996). *Ache Life History: The Ecology and Demography of a Foraging People.* New York: Hawthorne. ISBN 0-202-02037-1.

To be fair, The Selfish Gene is an excellent book and does correctly explain how evolution works most of the time at the genetic level. But unfortunately it is only half complete; taking Dawkin's and Darwin's and others' bottom up (inductive) approach from using the selfish gene as the starting point to explain all of evolution, leads them all into a blind alley from which there is no way out. The only way to discover the other missing half of the theory of evolution is to work the BIG PICTURE backwards from the top down to the gene (deductive reasoning) instead of trying to work the gene up to the BIG PICTURE.

If Dawkins and all the other Darwinians were completely right the world would be inhabited by Darwinian Monsters. So where are the monsters??

What is a Darwinian Monster? A "Darwinian monster" is a species/organism that will reproduce as soon as it is born by simply cloning itself and will never get old or die. It would be the perfect machine for making more and more copies of its selfish genes to eventually take over the world! One could imagine a renegade cancer cell that can exist outside of its victim and evolves the ability to engage in photosynthesis AND consume any biological matter it comes into contact with to make more copies of itself. If what Darwin and Dawkins promoted was all there was to evolution, that evolution was solely driven by the march of the selfish gene, such Darwinian monsters would not just be possible, they would be the norm! And human beings and most other life forms would not exist!

Okay, enough of Dawkins. I knew where he would be weak and clueless and it took about 2 minutes of looking in his index to see the failings of his and mainstream science's point of view in explaining aging, sex and menopause. So now let's get to the enjoyable stuff.

Let's construct and then SEE the BIG PICTURE

Chapter 1: The First Glimpse-Exceptionally Long-Lived Animals

I caught my first glimpse of the BIG PICTURE while studying the variations in aging/ life spans between species. It wasn't long after I began learning about aging that I encountered various charts showing the relationship between body size of animals and their life spans such as the following:

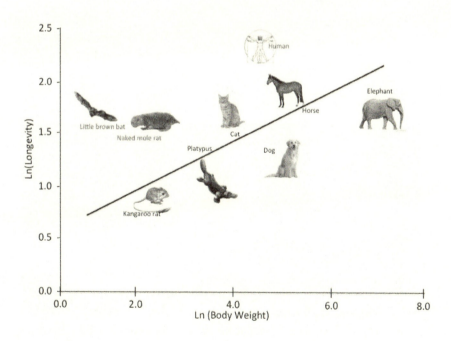

I started looking at these various charts and I noticed that most animals fell on an upward sloping straight line that related body size to life span. The general rule was that the bigger the body-size the longer the lifespan. So, for example, the line explains why a rat or a mouse can live only around 2 to 3 years, a dog about 10 to 15 years, a horse 40 years, and elephant 65 years. That was kind of interesting, but then I found what was more interesting was the small number of extreme outliers-animals that lived exceptionally longer lives than predicted by their body size. While the charts usually only included humans (120 years) bats (40 years) and some birds (90 years) as the exceptional long livers for their size, I started finding quite a few others that could be added as outliers to the body size/lifespan continuum.

Cookie, the oldest living parrot known, celebrates her 82nd birthday

A few humans have reached the age of 119 and one 122 years.

World's oldest wild bird raises a chick at age 63.

1951

1992

Here are two pictures of the Scottish ornithologist George Dunnet and a Fulmar. The first picture was taken in 1951 while Dunnet was tagging a young adult female bird. The young Dunnet has a full head of dark, wavy hair. The second picture, taken in 1992, shows the much older Dunnet while the fulmar remains indistinguishable from its 1951 picture. The Fulmar's secret?-the predation defenses of flight and extreme isolation. It lives mostly out on the open sea, rarely interrupting its flying to occasionally float on the water. It only returns to nest on very steep inaccessible cliff ledges. The Fulmar shows no signs of aging ever, and is thought to just drop dead at around the age of 60+ when its life has run its course. Like most long-lived species, Fulmars have a long period of development, 8 years to reach fertility, and have a small litter size, a single egg per year. And young chicks have evolved the predator defense trait of being able to shoot a very caustic oily substance from its gut with great accuracy.

This is the oldest bat living today-the male Brandt's **myotis** is at least 41 years old while weighing only about ¼ of an ounce.

-Small box tortoises can live 120 years or so but weigh less than most adult rabbits.

-Larger Galapagos tortoises have been known to live up to 190 years old.

While an Aldabra Giant Tortoise reached the ripe old age of 225 years old.

-Some lobsters have been dated to being more than 200 years old.

52 pound lobster largest and oldest ever caught in Maine-1926

-Arctic clams (**Arctica Islandica (Ocean quahog)**) have been known to live up to **500 years old** while being on the small side achieving a maximum shell height of 2 inches.

-Deep dwelling (isolated) fish like the Orange Roughy routinely live to 140 years old,

and the Rockfish can live up to 200 years.

-Isolated opossums living on predator-free islands can live up to 50% longer than opossums on the mainland and are found to be amazingly tame when encountered.

-Queen Ants are known to live up to 30 years,

Queen Bees 3 to 4 years which is about 40x the lifespan of the average worker bee.

Queen termites can live up to 50 years.

All of these queens live inside a protected hive or hill and are protected by their subjects. This makes the queens' living conditions similar to living within a cave (see next category).

-Cave animals all have dramatically longer lifespans than their above ground counterparts. Examples include: the blind salamander which can live 100 years while weighing just ½ an ounce. The blind albino cave crayfish has been reported to live to 176 years old whereas a normal crayfish only lives 3 to 8 years depending on the species.

Blind Cave Salamander

Blind Cave Crayfish

Similar to living in an isolated cave, there are many animals living miles beneath the ocean surface at the floor of the ocean. Now that is what I call isolation. And while not many species have been studied one of the few species they are able to date as to how old they are is the tube worm. Where some have been logged as being 800 years old!

If you look at the longest lived plants (generally trees) you will find most of them have evolved the same sort of predator defense strategy where there are no fungi, insects, plants or animals that eat them, and thus they can evolve very long life spans. What the defense to predation is for these long-livers is generally some sort of toxin in its bark, wood, leaves and seeds.

For example cedar is used to make chests that protect clothes because they repel moths and other pests. It just so happens that Japanese cedar trees can live up to 2300 years, while Northern White Cedar trees found in Canada have been known to be up to 1700 years old. Cedar mulch is used to repel insects from gardens as well as Cypress mulch. Do Cypress trees have long lives? The fourth oldest tree in the world is a Patagonian Cypress, 3644 years old in a national park in Chile. When looking for termite proof woods, there are a number of recommendations including Redwoods and Pine and eucalyptus. There exists on earth a 13,000 year old eucalyptus tree in Australia. There are various redwood trees that are up to 2200 years old (part of the sequoia family) and some American pines thought to be 1700 years old, and bristlecone pines are the oldest verified trees on earth living up to 5,100 years so far. Other woods that are recommended as tick repellents include Juniper trees' oil, Junipers are part of the Cypress family. Some Junipers in California are known to be at least 2,200 years old. All parts of the yew tree are toxic and the leaves are used to extract a drug used to treat cancer called taxol. Yew trees can live up to 5,000 years old. You get the picture; a great defense to predation allows an exceptionally long life span to evolve. Of all the defenses, poison has to be one of the best! If you are poison, then you won't be eaten!

13,000 Year Old Eucalyptus Tree-Australia

Bristle Cone Pine-5,100 Years old

Japanese Cedar Trees-2300 Years Old

Patagonian Cyprus 3644 years Old

Giant Seqouias (American Redwoods)- 2200 Years Old

California Juniper- 2200 Years Old

Yew Tree- 5000 Years Old

Now when you combine toxic poisons as a defense along with isolation and a harsh environment as multiple defenses to predation you can get some very old living organisms such as King Clone an 11,700 year old creosote bush (not even a tree) in the Mojave Desert. Creosote oil is toxic to fungi, insects, and marine borers and thus is used to treat wood for telephone poles and railroad ties and mine timbers. Because it induces vomiting in humans it has a medicinal use as an expectorant.

Creosote Bush- "King Clone" -11,700 Years Old

As interesting as the previously mentioned plants and animals are, there was one animal that really grabbed my attention: the desert-dwelling, underground-living, naked mole rat.

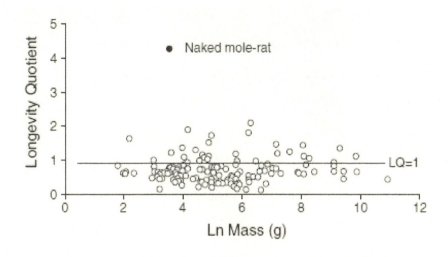

Relative life span of naked mole-rat to other rodents

While a normal rat grows old and dies at the age of 2 to 3 years, the similarly sized naked mole rat routinely lives to 28 to 30 years old. Why?

The naked mole-rat is well adapted for the limited availability of oxygen within the tunnels of its typical habitat; its lungs are very small and its blood has a very strong affinity for oxygen, increasing the efficiency of oxygen uptake. It has a very low respiration and metabolic rate for an animal of its size, about 2/3 that of a similarly sized rat, thus using oxygen minimally. In response to long periods of hunger, its metabolic rate can be reduced by up to 25 percent.

Naked mole-rats appear to have a high resistance to tumors; there are no known cases of cancer-infected naked mole rats.

Blind mole-rats *Spalax golani* and *Spalax judaei* also appear to be immune to cancer but by a different mechanism.

The naked mole-rat is also of interest because it is extraordinarily long-lived for a rodent of its size (up to 31 years) and holds the record for the longest living rodent. Naked mole-rats maintain healthy vascular function longer in their lifespan than shorter-living rats.

The most fascinating thing I found out about naked mole rats, however, is that in a colony of say 30 individuals, male and female, only one female reproduces and she is the queen. Also, only around 3 or so males mate with her, the rest of the colony basically consists of back-up individuals in case the queen or any of the princes die. What is so interesting is that apparently the queen emits some sort of signal that causes all the other females to remain infertile. When they remove infertile females from a queen's colony and put them into a new environment without a queen, they quickly morph into a fertile queen.

After studying all these different long-living animals <u>I finally figured out what they all had in common.</u> **They all had very effective defenses to being killed by predators**.

The defenses include

-full body armor: clams, lobsters, tortoises

-flight: birds, bats

-isolation: deep dwelling fish, desert animals, island opossums, naked mole rat, cave animals

-high intelligence: humans (Chimps and Gorillas also live a bit longer than they should, based on body size, so they probably should be included here too).

-living in a highly protected environment: cave animals, underground dwellers, colony queens, naked mole rat, bats

So the first piece of the puzzle of the BIG PICTURE is this

When a species evolves or acquires a very good defense to predation it allows exceptionally long life spans to evolve. The defense to predators allows the species to keep reproducing as long as physically possible, which over time selects for individuals who can live and reproduce at older and older ages.

Working backwards from this rule, we can deduce **that aging is selected for and conserved by evolution as a defense against evolving predation.**

This might be hard for you to swallow right now, let's just say for the time being, that it might be true. Once we fill in the rest of the BIG PICTURE we will find that many of the other puzzle pieces will also all point in this direction.

Some questions to answer:

How could aging be a defense to predation and predators?

Are these exceptionally long-lived animals descendants from even longer living animals that started aging in response to predation? OR

Were these long-living animals all just ancestors of shorter-living animals? OR

Do species' aging vary over evolutionary time frames in both directions depending on the environment?

CHAPTER 2: The Second Piece of the Puzzle – Rapidly Aging Animals.

After looking into animals with exceptionally long lives that age slowly, the next obvious area to investigate are the animals with exceptionally short lifespans for their body size.

This turned out to be a relatively under-researched area and I did find a chart that highlighted four animals with relatively short life spans based on their body size:

The flightless , large bird, the Emu (like an Ostrich): 19 years in the wild,
The Wallaby- like a kangaroo: 12 to 15 years,
The Papuan Forest shrew: 2 years maximum, and
The Pied Kingfisher: 4 years max., a water hunting bird that nests close to the ground and is the only Kingfisher species that also forages offshore.

These examples seemed to hint at the idea that exceptionally short-lived species might be more subject to predation-related deaths, but it was not much to hang one's hat on.

However, while looking into animals with extra short lives, I ran into a few remarkable animals and some plants that age rapidly and die at the same time they have a large one-time burst of reproduction. The class of animal that fits this description is known as a semelparous reproducer. A species is considered **semelparous** if it is characterized by a single reproductive episode before death. This in turn gave rise to the category aging researchers assign these animals to: semelparous aging.

What is interesting about modern evolutionary and aging theory is that semelparous aging is <u>considered a special category that exists all by itself</u> and **has nothing to do with other aging theory**. Why? Because even mainstream aging theorists can plainly see that this form of aging is programmed and not caused by gradual accidental damage.

However, luckily for me, when I first encountered semelparous aging organisms their existence just confirmed what I already knew in my gut, that aging is programmed , and that these semelparous agers are no exceptions to correct aging theory , but should prove to be a good starting point! From this category I found five very good examples: the Pacific Salmon, the marsupial mouse, the female octopus, annual plants, and bamboo trees.

Bamboo-First Flowering Right Before Death- Age 48 Years

Rotting Bamboo Forest (After Flowering)

Varous species of bamboo will flower at different ages, some take as long as 120 years, others 40, 65, or 80 years. Immediately after flowering and dropping seeds, the bamboo plant dies. No matter how far spread apart on the globe, all members of a single species of bamboo will flower and die at the same time. (Sounds a little like human menopause minus the flowering.)

After a little research I discovered the followng study that measured various hormone changes in the bamboo plant that immediately precede the flowering/aging/dying event. It turns out that **a few hormones increase while some others decline**. This will be shown to be VERY important later.

Guang Pu Xue Yu Guang Pu Fen Xi. 2013 Sep;33(9):2512-8.

[Responding relationship between endogenous hormones and hyperspectra reflectance for the different flowering stages of bamboos in Qinling].

[Article in Chinese]
Liu XH[1], Wu Y[2].
Author information
Abstract

The present research aims at utilizing hyperspectra information to estimate the content of endogenous hormones for predicting the bamboo flowering. The authors selected the bamboos Bashania fargesii, Fargesia ginlingensis and Fargesia dracocephala in different flowering situations to measure their hyperspectral reflectance in Foping Nature Reserve. The enzyme-linked immunosorbent assay was used to measure the contents of four endogenous hormones: Gibberellin (GA3), Auxin (IAA), Zeatin nucleotide (Zr) and Abscisic acid (ABA), and non-parametric test was applied to analyze their differences. The results showed that (1) GA3 contents of B. fargesii and F. dracocephala were higher in the flowering bamboo than in the non-flowering bamboo, while F. qinlingensis showed the opposite result; IAA and ABA contents of three bamboos were all lower in non-flowering bamboo than in the flowering one; Zr contents for three bamboo species were higher in the non-flowering bamboo than in the flowering one.

Annual Plants-Rapidly Die After Producing Seeds

The Marsupial Mouse

Marsupial Mouse Has Marathon Sex Until It Goes Blind and Drops Dead

Pacific Salmon-Ocean, River-Spawning, and 3 Days After Spawning

(Cigarette?   Was it good for you?)

The Brooding Female Octopus with Her Eggs

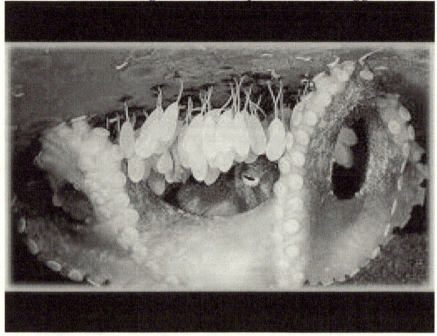

From Wikipedia:

"Female Octopus hummelincki lays eggs, broods them, reduces its food intake, and dies after the young hatch. **Removal of both optic glands** after spawning results in cessation of broodiness, resumption of feeding, increased growth, and **greatly extended life-span**. Optic gland secretions (aka as hormones-my note) may cause death of most cephalopods and may function to control population size.

The female octopus dies of starvation following brooding. This is evidence of a complex suicide mechanism involving the nervous system (behaviors). Some sort of suicide mechanism **triggered by reproduction** is signaling the nervous system such that the octopus does not feel hunger and therefore does not eat.

Experiments have shown that removing optical organs interferes with this behavior and results in an octopus that can survive reproduction. This in turn suggests that sensing of some external condition may play a part in the behavior of the octopus and also **demonstrates that the octopus does not die of "exhaustion" resulting from reproductive activity but rather as a result of a suicide mechanism**.

The most extreme experiments that demonstrate the cost of reproduction for longevity involve castration. **In annual plants, such as soybean, stripping the plant of flowers often prolongs life by months**. In animals that reproduce just once, castration before reproduction can increase longevity by years. **In Pacific salmon, it was possible to extend the life of a castrated fish by more than a decade (as opposed to a 3 year life span)**. In marsupial mice of the genus *Antechinus*, castrated males live months longer than intact males. There is some evidence that castrating institutionalized human males increases their life span. There are also some anecdotal reports that Korean eunuchs who were castrated to work with the king's harem tended to live and additional 10 to 15 years. The records of the British aristocracy also suggest that females who have fewer children also live longer, although much of this data antedates modern medicine. **All these examples indicate that reproduction affects longevity** within the lives of single organisms, but they do not indicate how reliable the effect is."

So I started thinking about WHAT did all these rapidly aging plants and animals have in common? For the most part, if they were all castrated they lived longer. (I did not find any information regarding preventing bamboo flowering when I looked into it. Not enough study has been done in many of these areas). Annual plants, if you pinch off their flower buds as they emerge, can live much longer than a plant that has "gone to seed".

I had the hunch at this point, that there was some sort of poison that was associated with the seeds, sperm and eggs that all these animals and plants were creating that was causing the plant or animal to rapidly age and die!

I tried to research the hormone changes in the various rapidly aging organisms and found in the 1990's there was not much to go on. All the aging theorists had decided that it was high levels of cortisol (the stress hormone) that increased dramatically and rapidly aged the Pacific Salmon.

There was no study that I could find where the reproduction-related hormones were investigated for their ability to also cause the rapid aging seen in these organisms!

For example, today if you do a google search of Pacific Salmon and hormones, this is about all you can find- a chart of their lifetime cortisol levels.

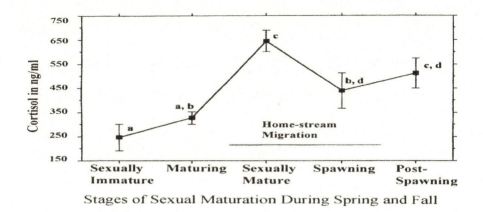

Stages of Sexual Maturation During Spring and Fall

Given that the rapid aging occurs post-spawning, I doubted that it was cortisol that was behind their rapid demise.

I then came up with the next possible rule of aging:

-Aging is driven by the action of reproductive hormones which have a 2-fold purpose:

1. To drive sexual development and reproduction , and

2. To cause the organism to age and die.

Now again, you do not have to accept this unusual idea as a rule just yet. It will become much more supported by the next glimpse of the BIG PICTURE- a study of how reproduction-related hormones vary with age in the most extensively studied organism in the world, the human being.

CHAPTER 3: The Third Puzzle Piece-Lifetime Human Reproductive Hormone Changes

(Lifetime Changes in the Human Reproduction –Related Hormones)

Let's start with melatonin. It is not normally thought of as a reproduction-related human hormone. You might know of it as a sleeping aid which is also good for jet lag in us humans. In many other animals, however, changes in melatonin levels control all aspects of mating behavior. Melatonin levels also control many other things like seasonal behaviors, the appearance of camouflage colors, etc. It works primarily by controlling the level of other hormones in the hormone (endocrine) system. Also, in human females, 75 mg of melatonin taken every night is an effective form of birth control. Let's see how it varies over a human's life span:

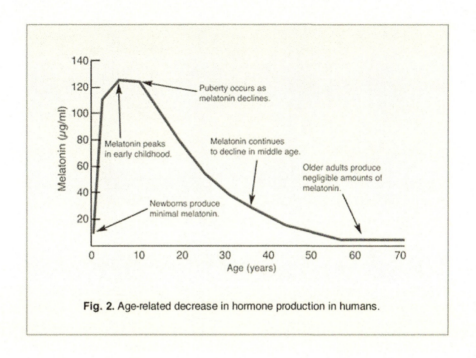

Fig. 2. Age-related decrease in hormone production in humans.

Now let's look at some less well known human reproductive hormones: Luteinizing Hormone (LH) and Follicle Stimulating Hormone (FSH). These are the hormones that kick in and drive the process of humans entering into puberty and when they increase in childhood they eventually stimulate production of higher levels of the more well-known reproductive hormones estrogen and testosterone.

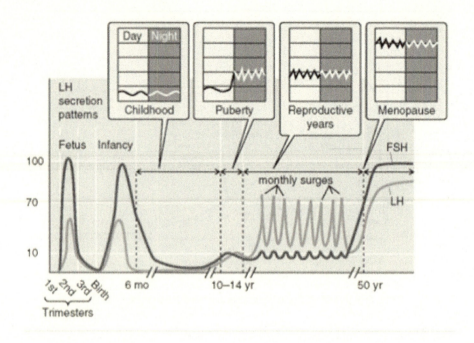

So you can see that FSH & LH levels vary dramatically over one's life time. In the fetus and the infant stage, these hormones increase dramatically and drive all sorts of development from an infant into a child. At about the age of 10, a new FSH & LH increase drives the development of the child into an adult by triggering puberty. Later, in mature females, monthly LH and FSH surges drive the development of immature eggs into mature eggs and their release into the fallopian tubes where they wait for fertilization.

Now here is the MAJOR CLUE that we are on the right track that reproductive hormones are also involved in driving aging: after age 50, FSH and LH increase dramatically back to the sky high levels only previously seen in rapidly developing infants. To the mainstream aging theorists this is just an accident of evolution with no particular purpose. From a

PROGRAMMED aging point of view, the dramatic LH and FSH increases around age 40 and 50 are driving the progam of menopause-the destruction of the woman's ability to have any more children, and then after age 50 it continues driving the process of destroying the woman's body, just at a slower rate than occurs in the rapidly aging animals.

A similar hormone pattern occurs in men at around the age of 50 as well-LH and FSH shoot way up to the levels they were at when the fetus and infant were undergoing rapid development and growth. It's just that in this case of the 50 year-old man, development equals destruction.

When viewed in this way one can say a day older is a day older in the human development program; it all is programmed from cradle to grave.

Now do you remember the pattern in the earlier chart presented? How it showed that melatonin declines dramtically with age. It just so happens that melatonin suppresses LH and FSH, and as we will see later is an excellent anti-aging hormone that extends the life span of mice by 20% and will most likely extend the lifespan of humans.

More interesting facts about LH and FSH are that when they rise too rapidly in young children it induces a problem called precocious puberty. Precocious puberty is treated by giving afflicted children injections of Lupron which suppresses their LH and FSH to about zero, and stops the premature puberty hormone signals.

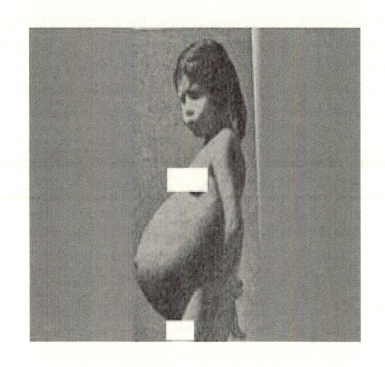

**LINA MEDINA,
A 5-YEAR-OLD PERUVIAN GIRL, WAS
THE YOUNGEST CONFIRMED MOTHER
IN MEDICAL HISTORY**

Melatonin and LH and FSH do their little dance over one's lifetime to first drive development from fetus to infant, then from infant to child, and then from child to teenager, and then from teenager to adult, and then from adult to elderly individuals, and eventually from elderly individuals to corpses.

LH and FSH and melatonin aren't the only hormones that show dramatic changes over one's lifetime. There are also a whole series of other hormones called steroids that also experience huge changes over one's lifespan. What is a steroid hormone? It is a hormone that your body makes from cholesterol as the initial building block. See the flow chart of how the steroid hormones are made by our bodies from cholesterol below:

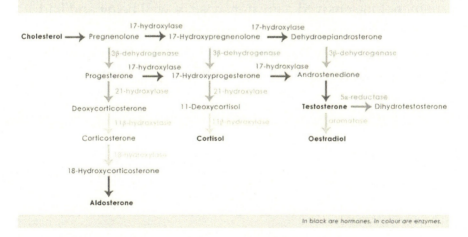

Many of these hormones decline dramtically with age, with the exception of the stress hormone cortisol (which tends to stay elevated as one ages).

For example consider the following hormone changes that occur with age:

Pregnenolone:

Pregnenolone is the first hormone that your body makes from cholesterol. It is also known as the grandmother of all other steroid hormones-Why? Because any steroid hormone found in your body was once at the beginning of its journey and transformation started off as pregnenolone. Pregnenolone is also known as the "Memory Hormone" as they have found when given to old rats and mice that they perform just as well as young mice on memory tasks. Pregnenolone is also given to humans to combat social shyness. At higher doses it has been used in the past to treat rheumatoid arthritis, and in those studies researchers found that those who took high dose pregnenolone enjoyed a greater sense of well being and had greater endurance as well. 400 mg a day was used in the studies. Notice how this wonderful hormone declines with age in both men and women.

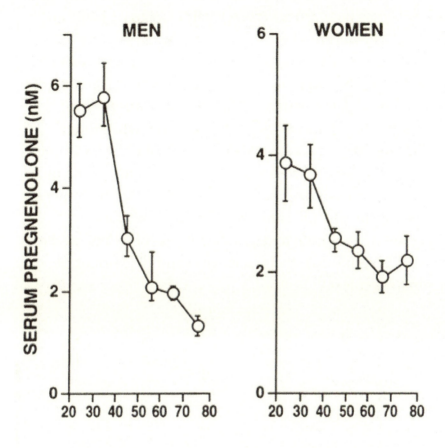

Progesterone: This is the second hormone made by your body from cholesterol. The cholesterol is converted to pregnenolone, and then the pregnenolone is converted to progesterone. It is said to be the most neuroprotective substance known to man. High levels of progesterone protect the brain and other nerves from injury and because women have much higher life time progesterone levels than men, they are known to recover from brain injuries much more quickly and better than men do. I have made the case in some of my prior books that progesterone administration will be found to be a good treatment for both Alzheimer's and ALS. The graph below shows how progesterone levels change with age in females. The decline is dramatic as progesterone drops to almost 0 after menopause. The change in progesterone in men is a bit different and I can find no charts for men's lifetime progesterone patterns (that shows you how unstudied this area is and how backwards we still are even while today's science seems to be so advanced!). But what I can say is that men's progesterone levels tend to increase little by little their whole lives on average, but then around the age of 70 mens' progesterone levels start to crash. I believe the crash is designed by evolution to begin the process of killing off the male. It is at this time that males' FSH and LH hormones begin to skyrocket!

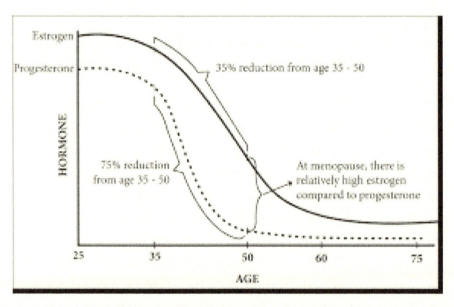

Lifetime Progesterone Levels-Women

DHEA: Another "miracle" hormone that declines dramatically with age is DHEA which is a well known antioxidant and muscle-building/ fat-burning anti-cancer/ antidepressant. It is really high in your system while you are a rapidly developing fetus; As a fetus it is about twice as high as your adult peak which occurs around age 25. After age 25 it is all downhill, heading towards 0 by the time you hit age 80. You can buy DHEA over the counter in the US but it is considered a prescription medication in Germany and is prescribed for depression.

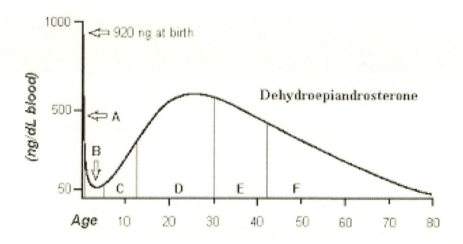

Testosterone: Another miracle hormone is the widely demonized testosterone, what we call "steroids" when we are referring to abuse by athletes. It declines dramatically with age in both men and women.

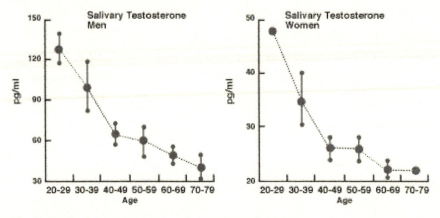

Fig. 3: Typical testosterone levels in men and women.

Testosterone peaks in humans at around age 20 and then starts a never ending decline:

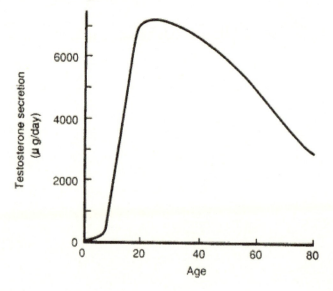

Estrogen: Estrogen levels decline with age in women as well but remain at relatively higher levels as compared to progesterone. In men the picture is a little murkier and estrogen in some men tends to increase with age suggesting to me that estrogen might be a sneaky little pro-aging hormone.

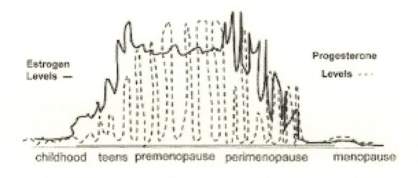

While we have been looking at lifetime patterns of various hormone changes, it is also interesting to note that these hormones also operate at the monthly level in women to control their fertility.

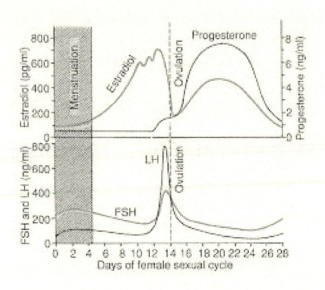

So basically we see that hormone levels in humans vary dramtically with age. It is very obvious to me and should be to you that they are behind the aging process. I will finish this chapter with various charts showing the levels and changes with age of hormones in both males and females. While not immediately obvious is that as a percent of baseline, LH tends to increase higher as percentage in women than men, and as a percent of baseline FSH levels have a higher percentage increase in males as compared to females. Because of this, I have dubbed LH as the female aging hormone, and FSH, the male aging hormone. And I can predict, and evidence is starting to support this idea, that diseases and consequences of aging more commonly seen in females will be found to be associated with LH, and those aging diseases and changes more commonly seen in men will be found to be associated with FSH.

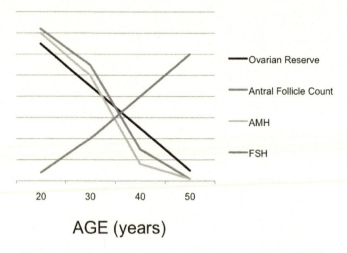

AGE (years)

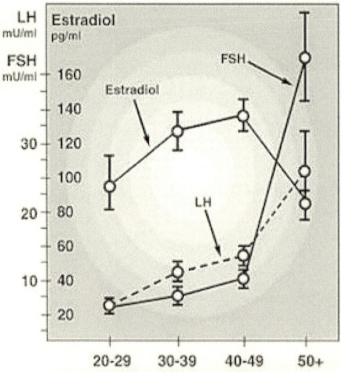

Figure 3. Alterations in FSH, LH, and estrogen with age. Note not only the alteration in magnitude of these hormones, but also the reversal of the FSH/LH ratio. Youthful FSH/LH ratios are less than "one." In the postmenopausal years, the ratio shifts to greater than "one" (Dilman and Dean, 1992).

Human Chorionic Gonadotropin: hCG is a much less well known reproduction related hormone than all the rest. It is quite similar to LH in its structure and size, and in fact hCG and LH can bind to the same hormone receptors. hCG however, in the female reproductive system, has a special function of causing the uterus to accept and nurture the fertilized egg rather than expel it as a foreign object. In fact, it is the fetus itself that secretes the large amounts of hCG during pregnancy. But it also turns out that men and non-pregnant women also make hCG their whole lives, but just at a lower level than a fetus during gestation. I had to search far and wide to find a study that measured hCG levels in men and women with age. Many scientists don't even realize that men can even make it! I was once at a science conference and was chatting with a cancer researcher from the esteemed Dana Farber cancer center who assured me that men could not make hCG. I was sure he was incorrect-and who was I to challenge such an eminent scientist? I looked it up...and Oh Yeah...men make it...I enjoyed watching his face turn red. Anyway as far as I know there was only one obscure study on the subject, and I was able to get the author of the study, Dr. Ulf from Finland to send me the following chart. And yes indeed hCG seems to also be a major aging hormone as you can see what happens on the chart as you age.

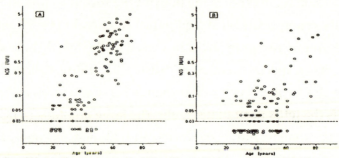

FIG. 2. Serum hCG levels in 106 nonpregnant women (A) and 100 men (B) as a function of age. ----, The detection limit of the assay.

TABLE 1. Serum hCG levels in men and nonpregnant women without evidence of malignant disease

	n	hCG (IU/L)			LH (IU/L)		
		Mean ± SD	Range	Median	Mean ± SD	Range	Median
Premenopausal women	54	0.15 ± 0.19	<0.03–1.0	0.05	7.04 ± 10.20	0.30–57	3.8
Postmenopausal women	52	1.41 ± 1.03^a	0.17–4.8	1.10	38.12 ± 16.18	14–75	37
Men <60 yr	83	0.18 ± 0.25	<0.03–1.3	0.04	4.13 ± 3.32	0.90–22	3.2
Men >60 yr	17	0.70 ± 0.76^b	<0.03–2.3	0.20	6.12 ± 2.94	2.6–13	5.1

For calculation of mean values, undetectable values were assigned a value of 0.03 IU/L.
^{a} $P < 0.001$.
^{b} $P < 0.01$.

concentrations of LH and hCG was less than 0.3%.
The recovery of hCG in the gel chromatographic fractions was 75–137% in four samples containing 2–5 IU/L hCG, as estimated by direct IFMA of serum samples. The addition of purified LH (13.5 mIU) to postmenopausal serum caused an expected increase in the LH peak from 2.5 to 5 IU/L, whereas the size of the hCG peak did not change (Fig. 4).

GnRH stimulation

Stimulation with GnRH caused a clear increase in serum hCG levels in all 13 subjects studied (Fig. 5). The mean increase was 3-fold in premenopausal and 2.5-fold in postmenopausal women ($P < 0.001$, by ANOVA). Peak levels occurred 30–60 min after GnRH injection. In 4 of the 7 premenopausal women and men, basal serum hCG levels were below the detection limit, but after GnRH administration, hCG was detectable in all subjects.

Estrogen-progestogen therapy

Therapy with estrogen and progestagen decreased the serum hCG levels in postmenopausal women ($P < 0.001$, by ANOVA and paired t test; Fig. 6). The changes in serum LH and FSH levels were similar to those in hCG. Three to 12 months after the beginning of therapy the mean serum hCG level had declined to 40–50% of the pretreatment level. However, in 4 of the 31 women, no significant change in hCG level occurred, although serum FSH levels decreased in every subject. In 2 of these women serum LH levels did not decrease either. In 2

The above graph showing the dramatic increases in hCG levels (a hormone similar to LH) is shown here courtesy of Dr. Ulf Stenman:

J Clin Endocrinol Metab. 1987 Apr;64(4):730-6.

Serum levels of human chorionic gonadotropin in nonpregnant women and men are modulated by gonadotropin-releasing hormone and sex steroids.

Stenman UH, Alfthan H, Ranta T, Vartiainen E, Jalkanen J, Seppälä M.

Growth Hormone: This declines with age as well but I don't believe it is an anti-aging hormone, but rather accelerates aging somewhat even though it boosts muscle mass and makes you look good. It creates a nice looking corpse if you take large amounts of it at older ages.

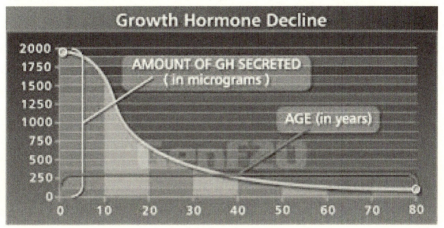

So the bottom line conclusion for this puzzle piece of the Big Picture is that in humans, just as in Pacific Salmon, and Flowering Bamboo, and Marsupial mice

-Human aging is driven by reproductive hormones that have what appear to be MULTIPLE roles:

1. To drive the development program from infancy to ADULTHOOD,

2. To drive the reproductive process during the reproductive years, and

3. To drive the aging process in the post-menopausal years.

Now this is quite a big pill to swallow? Is there any proof that this rule of aging is correct? Yes there is- some quite astonishing proof.

In 1998 I had my first science paper published in Medical Hypotheses a science journal out of the UK whose editors over the years had been winners of 5 Nobel prizes and three of whom had been knighted by the queen. Linus Pauling was once an editor, as well as Sir Karl Popper-a famous science theorist amongst fellow scientists.

In that paper, I predicted that increases in Luteinizing Hormone after age 50 in humans would be found to be involved in causing Alzheimer's disease. This was a simple prediction made from the observation that people who regularly took ibuprofen and smokers were less likely to get Alzheimer's disease than the normal population. It turned out the both ibuprofen and smoking suppress luteinizing hormone.

But this was a crazy prediction to make to most people, why? Because at the time luteinizing hormone receptors were thought to be only found in the sex related tissues. A few years after my prediction researchers found LH receptors were located all over the body including the brain! (In retrospect this shouldn't have been such an amazing insight given that LH increases are the first hormone changes that drive a child's body to morph into that of a teenager. LH obviously acted on the whole body and not just the gonads!)

Anyway, since that prediction in 1998 there have been quite a number of studies that have confirmed that yes indeed LH is involved in causing Alzheimer's and even the conservative US government run NIH has hopped onto the bandwagon.

I always like to fly this in the face of mainstream aging theorists who still insist that programmed aging is impossible. When they hear about the hormone that obviously drives the puberty/development program also being a hormone that drives the aging program and Alzheimer's they just go silent-crickets! They do not deny it; they just ignore it! In their case-ignorance is not bliss!

Since then it has been found that FSH, the other major reproductive hormone is involved in causing osteoporosis, and post-age-50 fat accumulation.

Eventually I expect they should find LH is also intimately involved in causing cancer and many other age-related illnesses.

CHAPTER 4: The Fourth Puzzle Piece- Asexual Versus Sexual Reproduction

There really is no bigger problem for the selfish gene viewpoint of Darwinian evolution than the existence of sex. Coincidentally, in the same month that my first journal article was published about the evolution of aging, the prestigious journal Science published a special issue about the Evolution of Sex. Here is the cover-

25 SEPTEMBER 1998 VOL 281, ISSUE 5385, PAGES 1913-209

Published the same month as my first paper>

Med Hypotheses. 1998 Sep;51(3):179-221.

The evolution of aging:

a new approach to an old problem of biology.

Bowles JT

I eagerly read all the articles in the journal and discovered for the first time that biologists had no idea why sex existed! It was a still a big mystery. There was one attempt by one theorist to suggest that the bright colors seen in various male bird species acted as a signal to the female about the health of the male's immune system!

That was bad enough, but the real problem they have with the existence of sex is that there exist a smaller number of organisms that do not need sex as a means of reproduction, rather they just clone themselves and pass on 100% of their selfish genes, rather than having to only pass on 50% which happens with sexual reproduction.

Examples of asexual organisms are limited, especially among animals. BUT THEY DO EXIST! Which means that all animals if their evolution was solely driven by the selfish gene trying to maximize copies of itself, should have evolved asexual reproduction long ago. Something just does not add up in Dawkins' and Darwin's world!

Asexual reproduction has been seen to occur in Hammerhead and Black Tip sharks, Boa Constrictors, whiptail lizards, some turkeys, and the Komodo Dragons!!

Parthenogenesis

- Produce eggs that develop into genetically identical versions of itself.
- The population is entirely female. The females lay unfertilized eggs that develop into genetically identical females.

The Whiptail lizard lives in the Sonoran Desert in Arizona. This lizard will lay an egg which will grow into another lizard.

NOW SOAK THIS IN!! THE IMPLICATIONS OF THIS ARE
HUGE!

Asexual Reproduction is not just some sort of weird extremely rare anomaly! It occurs quite often in very complicated animals! Which means it should be possible in ALL animals if not for various mutations that prevent it! The Selfish Gene theory should be championing these examples of proof of their theory's correctness!! Yet they ignore asexual reproduction! There is one entry in Dawkin's book "The Selfish Gene" in the index for asexual reproduction, and when you turn to the page, there is NO MENTION OF IT..for several pages before and after!!! More on this later…

And now, drum roll please, I have just found my SMOKING GUN! The missing link that pretty much proves all my top down ideas are correct-I give you the Brahminy Blind Snake:

Brahminy Blind Snake -Looks/Acts Very Much Like A Worm

The Brahminy Blind Snake may be the most successful asexual reproducer in the world, as it has spread to 6 continents. They are always female, never male. Why is it so successful? It is very small, 2 to 6 inches. It almost always lives underground and thus has no need for eyes. They are almost only encountered in the same way you encounter earthworms, during downpours where they are forced to surface for air, or when digging. This isolation, as we saw with the naked mole rat is a great defense against predators. However in the case of the Brahminy Blind Snake it led to the loss of sex as a defense against predation. I do not know if this snake is long lived- studies are needed. I expect it will also be long lived.

NOTE: I only encountered the case of the Brahminy Blind Snake while doing research for this book. To me this snake is the smoking gun-like Darwin's finches. I have always wondered why species that have escaped their predators for the most part, do not lose sexual reproduction, like they begin to lose their aging systems. Being focused on aging research for so long I never did much research into asexual animals, and it just remained a question in my mind. Looking at the few animals noted above that can reproduce asexually, most it seems are apex predators in their ecosystems like the Komodo Dragon, and Boa Constrictor, and large sharks. They all seem to be able to grow indefinitely with age. I expect they will all be found to have increasing fertility as they grow older. The one exception to these generalities seems to be the turkey! How could such an unremarkable bird that is not especially protected from predators more than any other bird, have evolved asexual reproduction!!?? Maybe there is a second path to the evolution of virgin birth; maybe species that have a really huge mortality rate in the males at times evolve asexual reproduction as a last resort? The male turkeys really do stand out, much like a peacock. Should we examine female peacock offspring for asexual female clones? This is an area for further research to figure out the underlying evolutionary dynamics. One prediction I can make, however, is that the existence of the Brahminy Blind Snake living so successfully and asexually underground suggests that many cave animals if studied closely enough will be found to be reproducing asexually. I did, however run a google search for asexual and cave animals and found that there exists a cave cricket has been found to reproduce exclusively asexually.

The Asexual Cave Cricket

Why are there no asexually reproducing mammals?

From: ferrebeekeeper.wordpress.com/2011/05/16/turkeys-and-parthenogenesis/

"Mammals do not naturally utilize parthenogenesis as a method of reproduction. Certain portions of mammalian genes consist of imprinted regions where portions of genetic data from one parent or the other are inactivated. Mammals born of parthenogenesis must therefore overcome the developmental abnormalities caused by having two sets of maternally imprinted genes. In normal circumstances this is impossible and embryos created by parthenogenesis are spontaneously rejected from the womb. Biology researchers have now found ways to surmount such obstacles **and a fatherless female mouse was successfully created in Tokyo in 2004**. With genetic tinkering, human parthenogenesis is also biologically feasible. Before his research was discredited and he was dismissed from his position, the South Korean (mad?) scientist Hwang Woo-Suk unknowingly created human embryos via parthenogenesis. To quote a news article by Chris Williams, "In the course of research, which culminated with false claims that stem cells had been extracted from a cloned human embryo, Hwang's team succeeded in extracting cells from eggs that had undergone parthenogenesis… The ability to extract embryonic stem cells produced by parthenogenesis means they will be genetically identical to the egg donor. The upshot is a supply of therapeutic cells for women which won't be rejected by their immune system, without the need for cloning."

All of which is fascinating to biology researchers (and those who would seek greatly prolonged life via biogenetic technologies), however it seems that in nature, the turkey is the most complicated creature capable of virgin birth."

And, finally, the stick insect of the genus Timema has been reproducing asexually for millions of years!

While it is somewhat rare, asexual reproduction occurs in enough large animals that <u>it cannot be ignored by a working theory of evolution</u> driven by the selfish gene! It is not even difficult to perform in the lab-as humans can make asexual clones by simply inserting a full set of DNA into a mammalian egg cell! If it is that easy for us, it must be easy for evolution to change sexual species to asexual in short order. YET asexual reproduction is almost completely ignored by the mainstream theorists. I say the widespread existence of sex is a fact that flies in the face of the idea of the selfish gene and is fatal to the idea that the selfish gene version of Darwinian evolution comes anywhere close to explaining what is seen in the real world.

All sorts of animals have been cloned:

But don't let me be too harsh on the selfish gene theorists, even though they ignore asexual species for the most part because they cannot explain the existence of sex, actually the existence of asexually reproducing females is the end result of selection driven by the selfish gene! So they ignore the examples of species that confirm their views because acknowledging these asexual Darwinian monsters would just showcase their ignorance and embarrassment on the subject of sexual reproduction.

The Reproductive "Cost" Of Sex

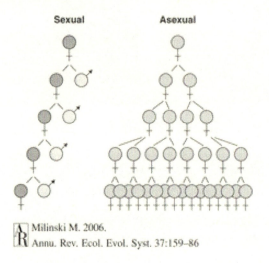

Milinski M. 2006.
Annu. Rev. Ecol. Evol. Syst. 37:159–86

From the point of view that sex and male/female sex types (to be discussed) are actually evolved defenses to evolving predation, it would make sense that every female of every species should be able to clone themselves when there is an absence of males as a last resort to prevent extinction of the species. I would predict that if you separate various females from males in many species, more than are known, that they may indeed give birth through parthenogenesis. As will be discussed later, in this BIG PICTURE way of looking at things, males are shown to have evolved to attract predators and to die in large numbers in order that the few males with predator tested genes can pass them onto the many camouflaged females. The risk of this strategy of course is what if all the males get eaten? It would simply make sense to then allow the females to reproduce clonally to prevent extinction of the group.

This of course suggests that females that can reproduce by cloning themselves should also have the ability to turn into males eventually when there are no other males around.

In fact there are quite a few species that have this ability which is called protogyny, here are just a few examples: the frog-<u>Rana temporaria</u>, , the star fish-<u>Asterina</u> pancerii, the sea slug-A. gibbosa,, the crustacean - <u>Heterotanais</u> oerstedi,, and many fish like the parrot fish, groupers, porgies, angel fish, gobies, emperors, wrasses, and swamp eels,

Female to Male Sex Changing Species

Female to Male Sex Changing Species

Female to Male Sex Changing Species

One thing one must notice about many of the previously shown animals that have females that can change into males, is that many seem to be very brightly colored. I am guessing that the brighter the color of the species, that the more likely the males are to all be eaten by predators , and thus the ability of surviving females to change into males becomes much more valued by evolution. This is surely an area ripe for study by future students of evolution.

Temperature changes have also been found to cause various Australian lizards to change sex such as the bearded Dragon lizard , being genetically male will develop as a female and even give birth to <u>more</u> lizards than if he/she were a genetic female. (This doesn't really pertain to the argument of females becoming males as a last resort, but it is interesting as another example of sex changing animals.)

From a 2011 article in Oxford's science magazine:

"It may come as a shock then, to realise that some of the best minds in the field have struggled to explain why sexual reproduction exists.

Given the benefits of sexual reproduction, the existence of asexual species seems odd. In fact, asexual reproduction allows the survival of species in extreme circumstances — for example, the female Komodo dragon will reproduce asexually when there are no potential mates.

However, in general, sexual reproduction is the rule and asexual reproduction is the exception. While there is not yet a generally accepted theory to explain this, it seems likely that the ability of sexually reproducing species to acquire multiple genetic advantages through recombination plays a key role."

I looked through the index in Richard Dawkin's The Selfish Gene book and saw one entry titled "asexuality" page 135. I turned to page 135 and read two behind , it , and two forward, ..No mention of asexuality..! This is to be expected as in Dawkin's world asexual reproduction should be widespread in the world, and in reality it is relatively rare. Dawkins won't even want to talk about it!

Now what I am going to share with you next might be a real shocker! At times human females also engage in the act of parthenogenesis. Human females from time to time attempt virgin births of asexually reproduced daughters. Unfortunately none of these have led to a successful birth. The only human virgin birth on record doesn't make sense since the mother had a son! When they do occur what happens is an unfertilized egg in the woman begins dividing and growing into a fetus. But science believes that there has never been a successful virgin birth among mammals. But who knows? Someday there might be enough mutations where it could happen. But in the meantime, doctors remove these little monsters from the pregnant mothers and are said to never show the mothers their little babies because they look so awful. Here are a few examples of failed human parthenogenesis-faces that only a mother could love!

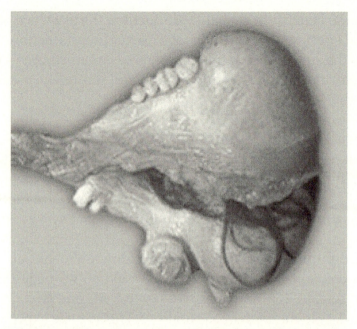

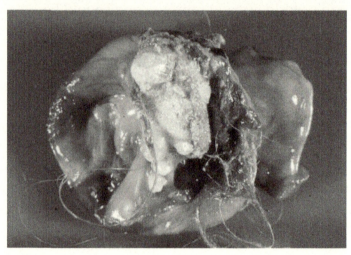

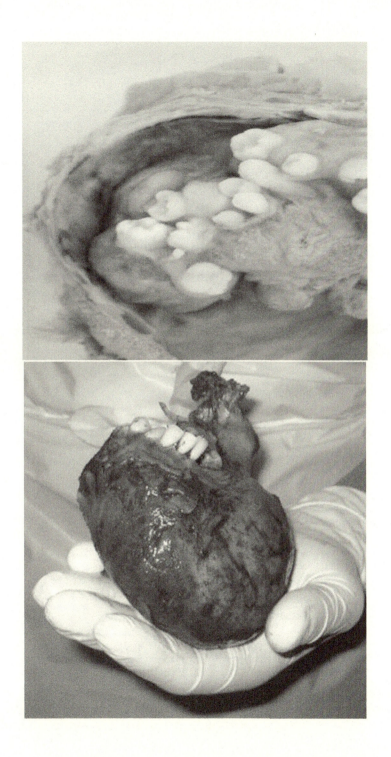

We are at the point where will try to explain sex within the framework that we have previously developed.

We have suggested that

-aging might be a defense to evolving predation, and

-that reproduction related hormones are associated with and also drive the aging process, and

-that a large burst of reproduction also results in a large burst of aging.

Where is all this leading us? To the idea that

<u>Sexual reproduction , like aging , is a defense against evolving predation!</u>

<u>Sex and aging are two sides to the same coin!</u>

While this could wait until later, let's just think about how sexual reproduction acts as a defense against evolving predation. Let's first consider a group of completely identical asexually reproducing animals. Say they are exposed over time to a number of different predators who are all evolving rapidly because they are sexual reproducers. The sexual reproducing predators will be quickly evolving better combinations of genes that can more effectively hunt and kill the identical prey species. So once the evolving predator has found a very good way to kill one of the prey clones, it has found a good way to kill them all-and thus the asexual prey species goes extinct in the evolving predator's parts of the world.

One way to look at it is as if a species defense to predation is like a combination to a safe. To just keep the single combination forever would be a quite risky strategy as the evolving predator is like a determined safe cracker always trying to solve the combination, and he has the advantage that he gets feedback. He can actually tell when he is getting close because he will start being more successful in his hunting. A species that can alter the combination quickly will be much more likely to defeat the safe cracking predator than a species that relies on a single combination for all time. Sex lets prey species change the combination quickly enough to survive.

Now there are other forces of death that can kill prey species such as starvation, or drought, or fires, or floods, but these forces of mortality are always the same, they never invent anything new, so after a long period of finding the best adaptation to these unevolving forces of mortality further change would be detrimental. Thus eventually, when confronted with solely unevolving forces of mortality, diversity would likely be detrimental, and thus, so would sexual reproduction as well as aging.

We might as well add aging to this example to get a grasp of the BIG PICTURE upon which we can build later. Aging simply seems to exist to limit the number of genes a single individual can contribute to the gene pool to maintain genetic diversity. If one successful prey animal could live forever she could wreak havoc on the gene pool's diversity and make the entire group more vulnerable to extinction by evolving predation. With this in mind we will see later that in species with two sexes, evolution concentrates much more heavily on limiting the female of the species' reproduction than the males. Why? Because in general females produce only a limited amount of offspring per lifetime and thus are the bottleneck through which population increases have to go through. The limit of females' reproductive potential is what limits the size of the population. Because of this, and because predator encounters cull a large number of males and demonstrates the value of older individuals' genes, the limiting of males' ability to reproduce is much less precise in exceptionally long lived animals.

A prime example of this is the declining fertility seen in humans. In females it is precisely controlled to occur around the age of 45 with very little variation in female menopause, while in human males the age at when reproduction is no longer possible varies widely primarily with the age of each male's death.

Darwin refers to the **survival of the fittest** as the main driver of evolution and in the short run that seems to be true. But in order to ever become the fittest, one also has to belong to a group that survives as well. So the larger picture driver of evolution in the face of evolving predation might be summed up as the **survival of the differing** referring to the group to which an individual belongs, and within that group a different form of selection occurs at the individual level which is Darwin's **survival of the fittest**.

Okay this is a really large leap and might be hard for some to swallow, but keep in mind we are just moving the puzzle pieces around and seeing how they might fit together. So rather than try to force this idea on you as it stands, let's look at another aspect of sexual reproduction , male and female sex types, and see if there is some way of looking at them where they can also be seen to be a defense to predation.

(One last thing-A confession: The last ovarian teratoma picture is actually a belt buckle made by someone with a strange sense of art. I just thought it had some good shock value so I added it for effect.)

CHAPTER 5: The Fifth Puzzle Piece- Male & Female Sex Types

As I said earlier, in Science's special issue about the evolution of sex published in 1998, the only idea put out that dared to try and explain the existence of sex was an article by a single brave contributor who theorized that the bright colors in males of various species were selected for by females because they demonstrated the health of his immune system! She was the only one who even tried to explain it. All the others just threw their hands up and admitted it was a mystery in the world of the selfish gene.

Well let's take a look at things that females of various species find attractive. I learned much of this by watching many, many, various nature shows on television by the way. Many females reject smaller males and only allow larger males to mate with them. This occurs in moose and sharks, vipers and lizards, sea mammals, etc. In many other species females select males with the brightest coloration. In other species the loud song of the male attracts the female. In other species the female selects the male with the longest tail, the biggest head, the largest antlers, etc. And there are males who are selected in other species who have the best dance moves! And then there are species where the female (usually) does not select at all but must mate with the male that is able to fight off all the other potential male suitors like elephants, and lions. (Female lions are also attracted to males that have the darkest and largest manes that stand out in the grasslands if they are allowed to choose).

What do all these male traits have in common? Let's recap

-Large body size

-Long tail length

-Most dramatic colors

-Loudest/best song

-Most elaborate dance

-Best fighter

-Biggest head

-Darkest mane

Think hard! If you think within the framework we are building of aging and sex as defenses to evolving predation it becomes pretty clear.

<u>All of these traits are indicative of the particular male's success in surviving encounters with predators!</u>

Most of these traits draw attention to the male from the female, like colors, songs, dances, dark manes in a light colored landscape etc. They not only attract females they also attract predator attention. If the male can survive a lifetime of running around with a target on his back, the female will be selecting predator tested genes.

As far as body size, tail length, antler size, head size go, these are all traits that display the age of the male. Why is this important? Because the older the male is, the more potential predator encounters he has survived. Let's take a look at some of the Big Pictures- I'll leave it to you to guess which one is the male and which is the female.

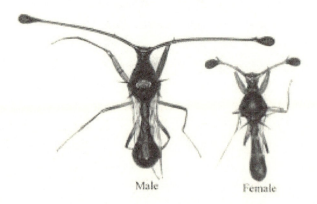

Teleopsis breviscopium

Male   Female

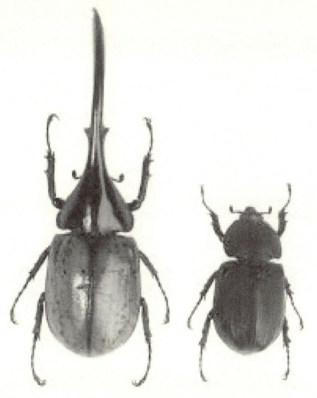

Do You Love Me? Now That I Can Dance?

Do You Love Me? Now That I Can Dance?

Beard And Moustache Length Indicate Age Of Male

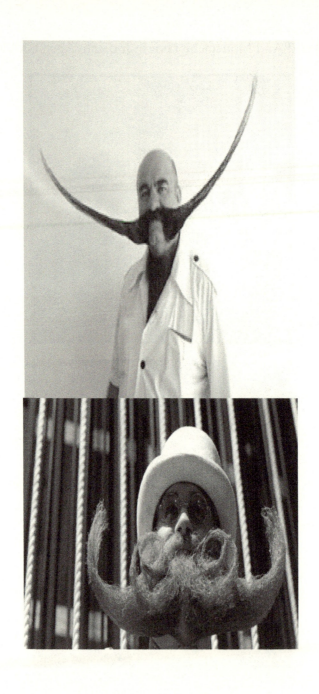

So as you might now have guessed, the male ornament for humans apparently is moustache and beard length as age markers. Many teenage boys can only grow a little peach fuzz sometimes until well into their 20's but beard length keeps getting longer and longer if uncut well into a male's late 30's. So beard length and moustache length are obvious sexually selected traits for human males. But are there any others? I believe so.

In addition to body size differences, probable female selected **male traits in humans that display age** and thus fitness in the face of predation are as follows:

-Larger Head/Nose

-Ear, Nose and Head growth

-Eyebrow growth

-Grey Hair

-Receding Hairline

So how do what we have learned about male and female sex types so far fit into the idea that sex as well as aging evolved as a defense to evolving predation?

In all animals with sex the only truly distinguishing thing between males and females is that males make sex cells by the millions that are quite small and only transmit genetic information. While in females they have a more limited number of larger sex cells (eggs) that transmit genetic information and also provide resources for the survival of the fertilized egg. Females are limited in their reproductive capacity while males basically have limitless reproductive potential.

(The above definition covers the male and female sex types of the more complicated life forms like animals. It also ignores for the moment much simpler life forms that do not have separate sex types like some single-cell organisms that do not have different sex types but still do engage in sexual reproduction by exchanging genetic material.)

Let's get back to the BIG PICTURE and figure out what is going on with the bigger than single-cell organisms. With respect to males, we see that they have essentially evolved to be the testers of predation. They are suited up with bright colors on their back, or forced to sing loud songs, or do crazy dances, all in an effort to attract a female. All the while **they are also attracting predators** who pick off the less fit males one by one, eventually leaving only the best survivors in the face of predation alive to escape to inseminate many of the more careful and camouflaged females with predator-tested genes. The beauty of this system is that if predators can kill 99% of the various males in an area, it only takes one survivor to pass on his predator tested genes to tens, or hundreds, dare I say thousands of females! I can think of no faster way to alter a local species' gene pool towards predator tested genes than this! As Mel Brooks once said: "It's good to be the king!"

So we have examined in this chapter, two major contradictions to the idea of the selfish gene. First, the existence of sex completely befuddles modern Darwinian evolutionists like Dawkins and almost all mainstream biologists, and always will. Passing on only 50% of your selfish genes instead of 100% is completely counter to what Darwinian evolution should predict. This is especially true after we have shown that asexual reproduction occurs enough in the animal kingdom in very advanced life forms that if the selfish gene really was in charge of evolution all the time, there is no question that all animals and even humans would reproduce asexually and the entire world would exist only of females.

This of course leads us to the second major contradiction to modern Darwinian evolutionary theory-the commonality amongst most male sex selected traits-a huge cost to survival fitness. Modern biologists are also dumbfounded by male sex traits. They only focus on why these male traits are attractive to females; they completely ignore the fact that bright colors, loud songs, and elaborate dances increase the odds that a predator will also be attracted!

It is easy to understand how these attractive male traits come not only at a large cost to requiring additional resources to develop on a male's body or in energy expenditures, but also at a large cost to the male's ability to survive in an environment full of predators hunting for their next meal. In the face of high levels of predation, various forms of camouflage evolve such as is shown in the following BIG PICTURES:

So we can obviously deduce that evolution has discovered camouflage as a survival trait, and liberally selects for it in many animals and most females of all species, but for some reason it does not give camouflage to male members of most species. Instead evolution endows the males with huge targets that a predator just cannot miss no matter how stupid the predator is! "Over here! Over here! " The male sex traits scream! There is no doubt that these male displays reduce their odds of survival in the wild in the presence of predation which is pretty much ubiquitous on the planet. And displays and colors aren't the only things attracting females and predators. Unfortunately in a book I cannot transmit sound. However if you walk through any woods anywhere in the world or even just listen to your backyard usually in the evening, you will hear many different birds, and insects, and amphibians, and various other organisms making loud sounds. Nothing is louder than the cicada that most people know, but there are also birds making sounds, locusts, frogs, etc.- everywhere you go on earth almost. Hardly any, and maybe none of these noise makers are female. They are all males- attracting females as well as predators.

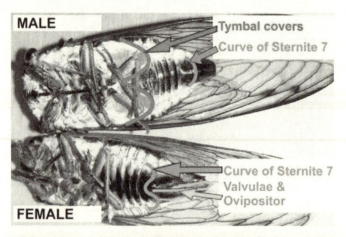

While male traits in many species involve bright colors and loud sounds and even behaviors (like dancing), additional attractive traits are more subtle. The subtle traits merely indicate the age of the male, and thus how many encounters with predators he has likely survived. Subtle male traits include size of various appendages such as head, ears, nose, length of tail, body size, etc. And in humans, as we have seen it includes beard length which if untrimmed, continues increasing in length well through a man's 30's.

And finally, modern evolutionary theorists in general believe that females of various species are only specifically attracted to the male trait seen in the males of their species. For example, it is widely assumed that say a female bird where males display blue wings will only be attracted to blue winged males. A very interesting study disproved this misconception.

Alexandra L. Basolo of the University of Nebraska found that when she attached artificial, plastic swords onto, naturally swordless platyfish males, female platyfish showed an immediate, strong and consistent preference for the males with the counterfeit swords.

This type of selection where females have preferences for male traits they have never been exposed to before has also been shown in Australian Finches by gluing various colored feathers to the males' heads. Females showed a dramatic preference for males with white feathers glued to their heads as opposed to the normal headed controls. For me this pretty much silences the argument that male traits are selected by females as anything other than that draws attention to the male, by her, and by predators.

Does any of this apply to humans?

CHAPTER 6: The Sixth Puzzle Piece- Homosexuality In

Animals & Humans

One would think that anyone defending the primacy of the selfish gene as the major driving force behind all of evolution would have some sort of reasonable explanation for how something as widespread as human homosexuality could evolve.

Homosexuality is a condition where the possessor of the homosexual trait will, if left to nature only, will choose to never have sex with the opposite sex and thus not have any offspring and not pass on a single gene to the gene pool! Yet human homosexuality exists and has persisted throughout history.

Certainly this must be harder for selfish gene promoters to explain than sex. At least with sex, the reproducer gets to pass on half of his or her genes. Here the selfish gene-ist has to explain how it evolved that someone passes on NO GENES WHATSOEVER! Given the very tough problem to solve here, the topic of homosexuality is just ignored for the most part, by evolutionary biologists.

Richard Dawkins , as courageous as he is, at least gives an explanation a try in a 2015 YouTube video. I share the link with you below. If you want a good laugh give it a watch. I am not laughing at Dawkins himself just at him trying to perform the impossible task of explaining homosexuality from the selfish gene point of view.

Darwin Day 2015 Questions: #4 How does evolution explain homosexuality?

Richard Dawkins Foundation for Reason & Science
https://youtu.be/IDmQns78FR8

After viewing this I think Dawkins might seem to focus more on male than female homosexuality and was so bold as to suggest that bottle feeding babies (male I presume) might make them more inclined to be homosexual. I am guessing he thinks sucking on a rubber nipple trains the young lad to want to suck on other protuberances. If nothing else, Richard Dawkins definitely has cojones!

Biologists in general tend to also discuss the evolutionary puzzle of homosexuality as mainly a human condition. Applying it to just humans makes the fact conveniently unique and a special category that can be ignored. They do it all the time. Menopause and suicide are also promoted by most biologists as being exceptions that apply to just humans.

Well it turns out that humans ain't the only ones with homos in the family!

There are some estimates that up to 1500 species have been documented to have homosexual individuals in their numbers! If you just do a quick perusal of the Wikipedia entry for homosexuality in animals you will get all sorts of examples. I will add their list here, deleting most of the text and leaving just the most amazing (or shocking) descriptions behind.

(from Wikipedia)

Some selected species and groups

Black swans

Swans, *Cygnus atratus*

An estimated one-quarter of all black swans pairings are of homosexual males. They steal nests, or form temporary threesomes with females to obtain eggs, driving away the female after she lays the eggs.[42][43] More of their cygnets survive to adulthood than those of different-sex pairs, possibly due to their superior ability to defend large portions of land. The same reasoning has been applied to male flamingo pairs raising chicks.[44][45]

Gulls

Studies have shown that 10 to 15 percent of female western gulls in some populations in the wild exhibit homosexual behavior.[46]

Ibises

Research has shown that the environmental pollutant methylmercury can increase the prevalence of homosexual behavior in male American white ibis. The endocrine blocking feature of mercury has been suggested as a possible cause of sexual disruption in other bird species.[47][48]

Mallards

Mallards form male-female pairs only until the female lays eggs, at which time the male leaves the female. Mallards have rates of male-male sexual activity that are unusually high for birds, in some cases, as high as 19% of all pairs in a population.[3] Kees Moeliker of the Natural History Museum Rotterdam has observed one male mallard engage in homosexual necrophilia.[49]

Penguins

Penguins have been observed to engage in homosexual behaviour since at least as early as 1911. George Murray Levick, who documented this behaviour in Adélie penguins at Cape Adare, described it as "depraved". The report was considered too shocking for public release at the time, and was suppressed. The only copies that were made available privately to researchers were translated into Greek, to prevent this knowledge becoming more widely known. The report was unearthed only a century later, and published in *Polar Record* in June 2012.[50]

In early February 2004 the *New York Times* reported that Roy and Silo, a male pair of chinstrap penguins in the Central Park Zoo in New York City had successfully hatched and fostered a female chick from a fertile egg they had been given to incubate.[12] Other penguins in New York zoos have also been reported to have formed same-sex pairs.[51][52]

Zoos in Japan and Germany have also documented homosexual male penguin couples.[30][31] The couples have been shown to build nests together and use a stone as a substitute for an egg. Researchers at Rikkyo University in Tokyo found 20 homosexual pairs at 16 major aquariums and zoos in Japan.

The Bremerhaven Zoo in Germany attempted to encourage reproduction of endangered Humboldt penguins by importing females from Sweden and separating three male pairs, but this was unsuccessful. The zoo's director said that the relationships were "too strong" between the homosexual pairs.[53] German gay groups protested at this attempt to break up the male-male pairs[54] but the zoo's director was reported as saying "We don't know whether the three male pairs are really homosexual or whether they have just bonded because of a shortage of females... nobody here wants to forcibly separate homosexual couples."[55]

Suki and Chupchikoni are two female African penguins that pair bonded at the Ramat Gan Safari in Israel. Chupchikoni was assumed to be male until her blood was tested.[59]

In 2014 Jumbs and Hurricane, two Humboldt penguins at Wingham Wildlife Park became the center of international media attention as two male penguins who had pair bonded a number of years earlier and then successfully hatched and reared an egg given to them as surrogate parents after the mother abandoned it half way through incubation.[60]

Vultures

In 1998 two male griffon vultures named Dashik and Yehuda, at the Jerusalem Biblical Zoo, engaged in "open and energetic sex" and built a nest. The keepers provided the couple with an artificial egg, which the two parents took turns incubating; and 45 days later, the zoo replaced the egg with a baby vulture. The two male vultures raised the chick together.[61] Two homosexual male vultures at the Allwetter Zoo in Muenster built a nest together, although they were picked on and their nest materials were often stolen by other vultures. They were eventually separated to try to promote breeding by placing one of them with female vultures, despite the protests of German homosexual groups.[63]

Pigeons

Both male and female pigeons sometimes exhibit homosexual behavior. In addition to sexual behavior, same-sex pigeon pairs will build nests, and hens will lay (infertile) eggs and attempt to incubate them.[*citation needed*]

Mammals

A female Labrador dog mounting another.

Amazon dolphin

The Amazon river dolphin or boto has been reported to form up in bands of 3–5 individuals enjoying group sex. The groups usually comprise young males and sometimes one or two females. Sex is often performed in non-reproductive ways, using snout, flippers and genital rubbing, without regard to gender.[64] In captivity, they have been observed to sometimes perform homosexual and heterosexual penetration of the blowhole, a hole homologous with the nostril of other mammals, making this the only known example of nasal sex in the animal kingdom.[64][65] The males will sometimes also perform sex with males from the tucuxi species, a type of small porpoise.[64]

American bison

The American Bison is a bovine mammal which displays homosexual behavior.

Courtship, mounting, and full <u>anal penetration</u> between bulls has been noted to occur among <u>American bison</u>. Could this be the origin of a word that describes a sex act that sounds like buffalo?
The <u>Mandan</u> nation Okipa festival concludes with a ceremonial enactment of this behavior, to "ensure the return of the buffalo in the coming season."[66] Also, mounting of one female by another (known as "bulling") is extremely common among <u>cattle</u>. The behaviour is hormone driven and synchronizes with the emergence of estrus (heat), particularly in the presence of a bull.

Bonobo and other apes

Two Female Bonobos

158

Bonobos, which have a matriarchal society, unusual among apes, are a fully bisexual species—both males and females engage in heterosexual and homosexual behavior, being noted for female-female homosexuality in particular. Roughly 60% of all bonobo sexual activity occurs between two or more females. While the homosexual bonding system in bonobos represents the highest frequency of homosexuality known in any species, homosexuality has been reported for all great apes (a group which includes humans), as well as a number of other primate species.

Bottlenose dolphins

Dolphins of several species engage in homosexual acts, though it is best studied in the bottlenose dolphins.[3] Sexual encounters between females take the shape of "beak-genital propulsion", where one female inserts her beak in the genital opening of the other while swimming gently forward.[77] Between males, homosexual behaviour includes rubbing of genitals against each other, which sometimes leads to the males swimming belly to belly, inserting the penis in the others genital slit and sometimes anus.[78]

Confrontations between flocks of bottlenose dolphins and the related species Atlantic spotted dolphin will sometimes lead to cross-species homosexual behaviour between the males rather than combat.[80]

Elephants

African and Asian males will engage in same-sex bonding and mounting. Such encounters are often associated with affectionate interactions, such as kissing, trunk intertwining, and placing trunks in each other's mouths. Male elephants, which often live apart from the general herd, often form "companionships", consisting of an older individual and one or sometimes two younger males with sexual behavior being an important part of the social dynamic.
Unlike heterosexual relations, which are always of a fleeting nature, the

relationships between males may last for years. The encounters are analogous to heterosexual bouts, one male often extending his trunk along the other's back and pushing forward with his tusks to signify his intention to mount. Same-sex relations are common and frequent in both sexes, with Asiatic elephants in captivity devoting roughly 45% of sexual encounters to same-sex activity.[81]

Giraffes

Male giraffes have been observed to engage in remarkably high frequencies of homosexual behavior. After aggressive "necking", it is common for two male giraffes to caress and court each other, leading up to mounting and climax. Such interactions between males have been found to be more frequent than heterosexual coupling.[82] In one study, up to 94% of observed mounting incidents took place between two males. The proportion of same sex activities varied between 30 and 75%, and at any given time one in twenty males were engaged in non-combative necking behavior with another male. Only 1% of same-sex mounting incidents occurred between females.[83]

Monkeys

Among monkeys Lionel Tiger and Robin Fox conducted a study on how Depo-Provera contraceptives lead to decreased male attraction to females.[84]

Japanese macaque

With the Japanese macaque, also known as the "snow monkey", same-sex relations are frequent, though rates vary between troops.

Lions

Male lions mating

Both male and female lions have been seen to interact homosexually.[86][87] Male lions pair-bond for a number of days and initiate homosexual activity with affectionate nuzzling and caressing, leading to mounting and thrusting. About 8% of mountings have been observed to occur with other males. **Pairings between females are held to be fairly common in captivity but have not been observed in the wild. (highlighting mine).**

Polecat

European polecats *Mustela putorius* were found to engage homosexually with non-sibling animals. Exclusive homosexuality with mounting and anal penetration in this solitary species serves no apparent adaptive function.[88]

Sheep

Ovis aries has attracted much attention due to the fact that around 8-10% of rams have an exclusive homosexual orientation. Furthermore, around 18-22% of rams are bisexual.[90]

An October 2003 study by Dr. Charles E. Roselli et al. (Oregon Health and Science University) states that homosexuality in male sheep (found in 8% of rams) is associated with a region in the rams' brains which the authors call the "ovine Sexually Dimorphic Nucleus" (oSDN) which is half the size of the corresponding region in heterosexual male sheep.[32]

The Merck Manual of Veterinary Medicine appears to consider homosexuality among sheep as a routine occurrence and an issue to be dealt with as a problem of animal husbandry.[93]

Spotted hyena

The spotted hyena is a moderately large, terrestrial carnivore native to Africa.

The family structure of the spotted hyena is matriarchal, and dominance relationships with strong sexual elements are routinely observed between related females. Due largely to the female spotted hyena's unique urogenital system, which looks more like a penis rather than a vagina, early naturalists thought hyenas were hermaphroditic males who commonly practiced homosexuality.[94][not in citation given] Early writings such as Ovid's *Metamorphoses* and the *Physiologus* suggested that the hyena continually changed its sex and nature from male to female and back again. In *Paedagogus*, Clement of Alexandria noted that the hyena (along with the hare) was "quite obsessed with sexual intercourse." Many Europeans associated the hyena with sexual deformity, prostitution, deviant sexual behavior, and even witchcraft.

The reality behind the confusing reports is the sexually aggressive behavior between the females, including mounting between females. Research has shown that "in contrast to most other female mammals, female *Crocuta* are male-like in appearance, larger than males, and substantially more aggressive,"[95] and they have "been masculinized without being defeminized."[94][*not in citation given*]

Study of this unique genitalia and aggressive behavior in the female hyena has led to the understanding that more aggressive females are better able to compete for resources, including food and mating partners.[94][96] Research has shown that "elevated levels of testosterone in utero"[97] contribute to extra aggressiveness; both males and females mount members of both the same and opposite sex,[97][98] who in turn are possibly acting more submissive because of lower levels of testosterone in utero.[95]

Others

Lizards

Whiptail lizard (Teiidae genus) females have the ability to reproduce through parthenogenesis and as such males are rare and sexual breeding non-standard.[99] Females engage in sexual behavior to stimulate ovulation, with their behavior following their hormonal cycles; during low levels of estrogen, these (female) lizards engage in "masculine" sexual roles. Those animals with currently high estrogen levels assume "feminine" sexual roles.

Lizards that perform the courtship ritual have greater fertility than those kept in isolation due to an increase in hormones triggered by the sexual behaviors. So, even though asexual whiptail lizards populations lack males, sexual stimuli still increase reproductive success.

From an evolutionary standpoint, these females are passing their full genetic code to all of their offspring (rather than the 50% of genes that would be passed in sexual reproduction). Certain species of gecko also reproduce by parthenogenesis.[100]

Insects and arachnids

There is evidence of homosexual behavior in at least 110 species of insects and arachnids.[101] Homosexual behavior in insects has been a matter of significant debate among scientists. Though various theories have been proposed, the leading explanation for why males, in some species up to 85%, engage in homosexuality is simply a case of mistaken identity. The cost of missing an opportunity to copulate with a female is greater than the risk of engaging in sexual activity with an insect or spider of the same sex.

Dragonflies

Dragonfly (*Basiaeschna janata*)

Male homosexuality has been inferred in several species of dragonflies (the orderOdonata). The cloacal pinchers of male damselflies and dragonflies inflict characteristic head damage to females during sex. A survey of 11 species of damsel and dragonflies has revealed such mating damages in 20 to 80% of the males too, indicating a fairly high occurrence of sexual coupling between males.

Fruit flies

Male _Drosophila melanogaster_ flies bearing two copies of a mutant allele in the fruitless gene court and attempt to mate exclusively with other males.[21] The genetic basis of animal homosexuality has been studied in the fly _Drosophila melanogaster_.[105] Here, multiple genes have been identified that can cause homosexual courtship and mating.[106] These genes are thought to control behavior through pheromones as well as altering the structure of the animal's brains.[107][108] These studies have also investigated the influence of environment on the likelihood of flies displaying homosexual behavior. (end Wikipedia)

So apparently homosexuality cannot be dumped into the unique human exception category and thus can no longer be ignored by biologists. It must be addressed; trying to address it from the perspective of the preeminence of the selfish gene just leads us into another blind alley with no way out.

What the Wikipedia entry fails to describe are the conditions affecting the pregnant mother of future homosexual offspring.

I'll just spill the beans right now. A number of studies in rodents have shown that if you stress the pregnant mother at certain times during her pregnancy she will tend to give birth to homosexual males and promiscuous females and a smaller number of homosexual females.

What is this telling is from the perspective of the BIG PICTURE? What causes stress? Too many close encounters with predators. This fits quite easily into the BIG PICURE of most unexplained biological phenomenon as being defenses to evolving predation.

How is having homosexual offspring a defense to predation? Having homosexual offspring is a form of birth control for mothers who are considered by evolution to be unfit in the presence of predation. The stress from predator encounters if extreme enough can kill the mothers and their unborn babies. If the stress is less extreme it can lead to homosexual offspring that need to be nursed for a relatively significant period of time. Nursing prevents the mother from becoming fertile for mating. So in a sense homosexual offspring is just nature's form of birth control for unfit mothers.

Let us consider the case of female offspring of stressed mothers being more promiscuous than the female offspring of non-stressed mothers. This also jibes well with the BIG PICTURE as promiscuous females who have offspring from multiple males rather than bonding with a single one will add more diversity to the gene pool than if she just mated with a single male for life. As we will see later diversity in the gene pool is the defense to evolving predation that evolution seeks with all these mysterious adaptations.

Now we get to humans; is there any evidence that stressing pregnant female humans can cause their male offspring to be born as homosexual? I wrote about this topic in my December 2000 paper published in Medical Hypotheses titled "Sex, Kings, & Serial Killers and other Group Selected Traits"

Here is the excerpt:

Homosexuality: Birth Control For Unfit Mothers?

Prevailing evolutionary theory cannot explain the conundrum of homosexuality. Current theory requires defining homosexuality as an evolutionary accident as homosexual offspring would not be expected to reproduce. Is evolution so sloppy that the sexual preferences of 10% to 20% of the human population **(78)** is simply a random mistake of nature? And why does it also occur throughout the animal kingdom from sheep **(79)** on down to rats **(80)**? If one accepts group selection as a reality, the purpose of homosexuality has a simple explanation.

Various studies show that when stressed at a certain times during gestation, rats give birth to males that exhibit female behavior and females that are more masculine **(81)**.(The literature is relatively conclusive on this for males, but the data on females is somewhat ambiguous. Some female offspring of stressed rats also show more promiscuous mating behavior). Stress increases cortisol levels in rats, and the Prior Paper referred to studies showing that cortisol appears to oppositely affect the sex hormones in human females and males which we will assume extends to rats.

If stress induces high maternal cortisol levels during gestation and the cortisol reaches the developing embryo, endogenous embryonic sex hormones may be altered. Testosterone and estradiol levels in male and female embryos respectively may be decreased. Decreased embryonic sex hormones likely affect the development of the brain's sexuality. It has been shown that the prenatal stress-induced feminization of male rats is prevented by perinatal androgen treatment **(82).**

Studies have shown that human females, male transsexuals, and homosexuals share similarities in certain brain structures which differ with heterosexual males **(83, 84)**. Also, it is believed that testosterone derived DHT is required during fetal brain development to create a "male brain"**(85)**. Likewise we might assume that estradiol exposure creates a "female brain" by feminizing some brain structures. If a stress-induced maternal cortisol surge suppresses the embryo's testosterone or estradiol, then homosexual offspring, of either sex could result. Interestingly, some researchers found that in a large group of homosexuals interviewed in Germany, many more were born during the war years of 1941 to 1947 than before or after this stressful period with the birth peak occurring in 1944-1945 **(86)**.

Why would evolution create such a system? If a pregnant female is stressed in the wild, it may imply close encounters with predators or maladaptation to her group. Evolution, through group selection, has likely selected for groups that remove or inhibit the spread of her "less fit" genes. While a spontaneous miscarriage or stressed-induced cannibalization of her young (which is common in rodents) is a simple solution, it would leave the female ready to reproduce again. A more clever and effective solution is to give her effectively sterile offspring which she will raise, and which will keep her from reproducing much longer than if she were childless. Also, if group survival required the homosexual children to reproduce, homosexual females could be forced to have sex by dominant heterosexual males. Homosexual males, however, who could not be forced, are evolutionarily irrelevant anyway as long as a single heterosexual male existed.

The only attempt at an evolutionary explanation of homosexuality that the author could find was one that proposed that a homosexual male child would be generated if it was prenatally stressed. The stressor was assumed to be the mother's living in a crowded environment. The homosexual male, as an adult would not reproduce so that in times of famine there would be fewer grandchildren, and thus an increased likelihood of the grandchildren's survival **(87)**. One does not have to work long to find counter arguments to this reasoning, but it is a creative attempt to overcome the conundrum of homosexuality and borders on using group selection as an argument. It is only referenced here to show the difficulties that exist in trying to explain homosexuality without the unabashed acceptance of some form of group selection (which we will later find to be a higher level form of selection called species selection).

One must wonder about the seemingly high levels of human homosexuality. Were so many mothers severely stressed by predators or wars during pregnancy? Not likely. However, a source of artificial stress has been unleashed this century on humans in epidemic proportions: cigarette smoking. Nicotine from smoking induces a significant increase in cortisol levels. If a pregnant female has the genetic predisposition to bear homosexual children when stressed, and she smokes during early pregnancy, the nicotine-induced cortisol increase may be sufficient to induce homosexuality in her offspring. This speculation could easily be confirmed or refuted with a simple epidemiological study.

78. Sell R. Wells J. Wypij D. The prevalence of homosexual behavior and attraction in the United States, the United Kingdom, and France: results of national population-based samples. Archives of Sexual Behavior 24(3). 1995. 235-248.

79. Perkins A. Fitzgerald J. Moss G. A comparison of LH secretion and brain estradiol receptors in heterosexual and homosexual rams and female sheep. Hormones & Behavior 29(1). 1995. 31-41.

80. Ferguson T. Alternative sexualities in evolution. Evolutionary Theory 11(1). 1995. 55-64

81. Ohkawa T. Sexual differentiation of social play and copulatory behavior in prenatally stressed male and

female offspring of the rat: the influence of simultaneous treatment by tyrosine during exposure to prenatal stress. Nippon Naibunpi Gakkai Zasshi-Folia Endocrinolgica Japonica. 63(7):823-35, 1987 Jul.

82. Dorner G. Gotz F. Docke W. Prevention of demasculization and feminization of the brain in prenatally stressed male rats by perinatal androgen treatment. Experimental & Clinical Endocrinology. 81(1):88-90 1983 Jan.

83. Swaab D. Gooren L. Hofman M. Gender and sexual orientation in relation to hypothalamic structures. Hormone Research. 38 Suppl 2:51-61, 1992.

84. Zhou J, Hofman M. Gooren L. Swaab D. A sex difference in the human brain and its relation to transsexuality. Nature. 378(6552):68-70, 1995 Nov.

85. Connolly P. Choate J. Resko J. Effects of endogenous androgen on brain androgen receptors of the fetal rhesus monkey. Neuroendocrinology. 59(3):27 1994 Mar.

86. Dorner G. et.al. Prenatal stress as possible aetiogenetic factor of homosexuality in human males. Endokrinologie.75(3):365-8, 1980 Jun.

87. ibid. 81.

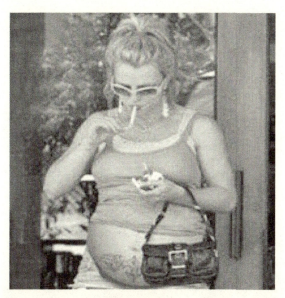

While searching for the old studies that showed a sharp rise in the birth of homosexuals in Germany during the WWII years (this supposedly also happened in England as well amongst the pregnant women who hid in London's subway tunnels during the German bombing campaign s) I found an article about a new book by Dr. Dick Swaab a well-known neuroscientist who is best known for his research and discoveries in the field of brain anatomy and physiology, in particular the impact that various hormonal and biochemical factors in the womb have on brain development. The book is called "We Are Our Brains" and he puts forth the controversial "new" idea that homosexuality occurs in the brains of fetuses in the womb of stressed mothers. He also notes , like I did in my 2000 paper, that smoking by pregnant mothers can lead to homosexual offspring because nicotine stimulates the release of the stress hormone cortisol and in effect acts as a predator encounter as perceived by evolution. He does add some new evidence that stress in mothers causes homosexuality by noting, as one would expect, that amphetamines (also fake stress) and various other substances lead to an excess of homosexual offspring.

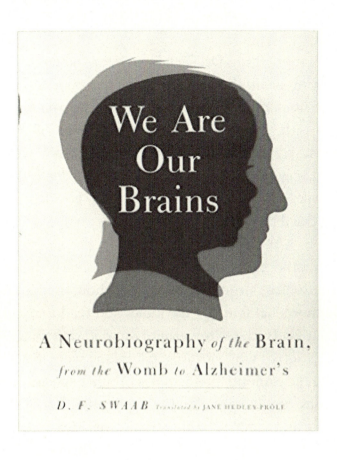

So that's about it for homosexuality. If we have learned another fact about evolution we can say that homosexuality fits in to the BIG PICTURE as:

1. Homosexuality is birth control for mothers perceived as stressed by evolution and thus possibly having less than optimal genes for the particular environment. Having an effectively "sterile" child reduces her potential contribution to the gene pool by preventing her from getting pregnant while nursing, and investing resources into the sterile child which also reduces her total potential reproductive output.

One more thing-Dr. Dick Swaab for some reason is getting death threats from some gay people who don't like homosexuality being portrayed as a pathology! I guess they want to think it is a choice. But if you ask most gay people they are happy to say they were born that way. So please don't send me any death threats thank you. Wait, on second thought, go ahead, it might help sell the book!

One other thing I have seemed to notice is that when you look at large groups of either homosexual males, or homosexual females, the males still tend to maintain their evolved desire to stand out and draw attention to themselves while groups of lesbians seem to act more like the camouflaged females of other species, who desire and have evolved to avoid attention. A straight male and female couple can test this idea by trying to enter a gay bar or a lesbian bar for a drink and see what happens! You can also test this idea by searching google-images for group of gay men and then for group of lesbians and you will see the difference! All the pictures are like what follows:

CHAPTER 7: The 7th Puzzle Piece- Human Menopause & Age-Changes In Fertility In Various Animals

Another **major mystery** for the mainstream "the selfish gene is the only game in town" lobby is the existence of **human menopause**. What is obvious to the selfish gene-ists' point of view and to me and anyone else is that menopause evolved to stop the human female from reproducing at around age 45 or so. Where the disagreement starts is why? The only thing selfish gene theorists have to hang their hat on is what they call "the grandmother hypothesis". Which to me is ridiculous on its face, I don't even have to do the math, it just has to be plain wrong. They are trying to get us to believe that a menopausal woman's genes are spread and survive better if she quits reproducing at age 45 and then spends the rest of her life helping to raise her grandchildren and improving their chances of survival. So instead of continuing to contribute 50% of her genes to each child she has after age 45, they suggest that evolution has stopped her from reproducing because her time would be better spent using her time to help her grandchildren to survive who share just $1/4^{th}$ of her genes.

Because the maximum lifespan of human females is 122 so far recorded to date, the implicit assumption here is that ancient human women used to be able to reproduce until possibly age 122! Then, in order to maximize the spread of her genes evolution has shut off her reproduction at age 45 so she can switch from reproduction to looking after her grandchildren. But for some reason evolution forgot to stop men from reproducing after age 45 so they could also help their grandchildren. The whole thing is ridiculous, ignores the life histories of the rest of the animal world and has been shown to be incorrect by a detailed study of a foraging tribe that lives in the jungle, the Ache people of Paraguay, which are representative of what small hunter gatherer tribes of people used to be like in the distant evolutionary past. (See <u>Ache life history: The ecology and demography of a foraging people</u> KR Hill, AM Hurtado,Transaction Publishers).

Here is the Wikipedia entry for evidence against the grandmother hypothesis:

Evidence against

Such historical studies are, however, unable to quantify grandmotherly assistance; they are merely correlations between infant mortality and the existence of a grandparent. One study that calculated grandmaternal assistance to both offspring and grandchildren did not find appreciable effects to warrant termination of fertility as early as 50.[25]

Another problem concerning the grandmother hypothesis is that it requires a history of female philopatry. Though some studies suggest that hunter-gatherer societies are patriarchal, mounting evidence shows that residence is fluid among hunter-gatherers and that married women in at least one patrilineal society visit their kin during times when kin-based support can be especially beneficial to a woman's reproductive success.

Others dispute the hypothesis, arguing that the grandmother herself will use up resources that could be used for new young.

In addition, all variations on the mother, or grandmother effect, fail to explain longevity with continued spermatogenesis in males. It also fails to explain the detrimental effects of losing ovarian follicular activity, such as osteoporosis, osteoarthritis, Alzheimer's disease and coronary artery disease. (end Wikipedia)

So, if the grandmother hypothesis cannot be supported, the selfish gene theorists are basically, again, left high and dry after the tide goes out, swimming naked for all to see!

Let us try to look at the mystery of menopause from the BIG PICTURE perspective which is always driven by an evolving predator.

Let us start with what is normal in normally aging animals which are the animals that can be found tightly hugging the body size/maximum lifespan trade off line. You remember the graph, here it is just for a refresher:

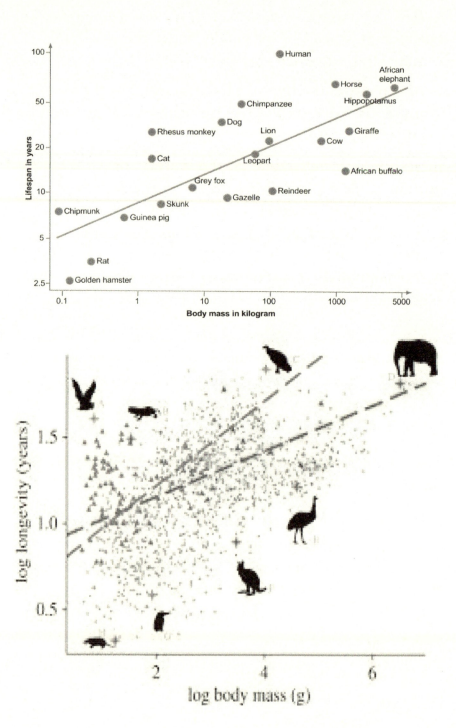

What is true about almost all land animals on the line or below the line like mice, rats, lions, dogs, skunks etc. is that as the female gets older, her litter size gets smaller and smaller until it finally reaches 0. It is usually at this age that she dies and males of her age also die.

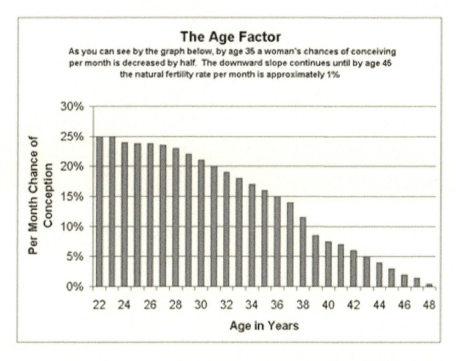

Above is the human female's chance of getting pregnant based on her age,

this basically shows a gradual reduction of litter size from 1 to 0 at the population level.

Below is a graph of various mouse strains and the maternal age and litter size as an example:

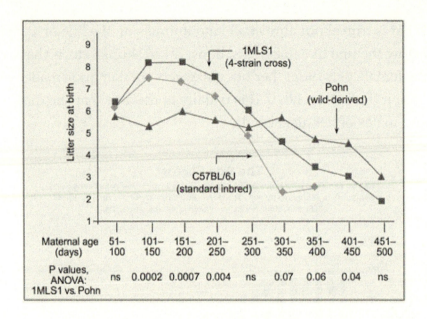

Also as female mice age the likelihood of them getting pregnant again declines (the same as in humans) over time as shown below for some various strains:

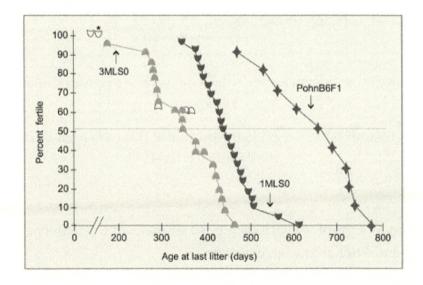

I just grabbed a random species' chart for female age and litter size and could not find a dog chart on the internet but did find a hedgehog graph as shown below:

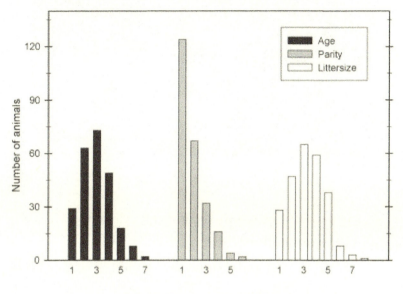

And the Norway Rat (Below)

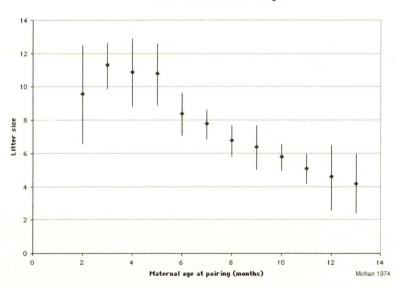

The logic is usually the same, litter size shrinks to zero at about the time of death of the female from old age. This is also usually the age at which the male dies from old age as well.

So how could this have any relevance to human menopause, where the female can live as long as 70 years past the age of about 45 or 50 where litter size shrinks from 1 to 0?

If our human ancestors originally had life histories similar to most other animals, female and male humans would both have lived at most to about age 45 or 50. Around this age females would lose the ability to reproduce, and both female and male would quickly drop dead. Later in this chapter we will see that the hormonal changes that caused the abrupt menopause (hormonal attack upon the female reproductive organs and capacity) also likely caused both male and female to come to an abrupt end by an incidental attack on the body.

However, over the ages, a human female here or there evolved a hormonal pattern altered enough that she could live a few years after her cessation of being fertile. What this did was opened the door to the evolution of longer and longer post-menopausal life spans. How is this possible? After menopause the female cannot pass her genes on, so any evolved life span extension she could muster could not be passed on to her children! If it was up to just her, post-menopausal life span could not evolve. So how did it happen?

The females that evolved postmenopausal life spans caused by altered hormone patterns, passed the same life span extension onto their sons! (As we saw earlier, the same hormones that increase dramatically with menopause in the female also see a similar rise in males starting around age 50 as well). And what did this accomplish? It allowed the sons to continue living longer and having more offspring. This is how post-menopausal life span in humans must have evolved and became greatly extended compared to normal animals. I call it **the son-king hypothesis.** If a female could live longer due to various mutations in her lifetime hormone patterns she would pass these off to her sons. If her sons became a king with a harem, her extended post-menopausal life span would be passed on to her grandsons and granddaughters. This didn't do the daughters any good from an evolutionary perspective as they still were menopausal at the evolutionarily tightly controlled age range of 45-50. However, it allowed her son-king to continue reproducing like a rabbit into much older ages.

The proof of this hypothesis lies in the genes of most Asian men.

From Wikipedia:

Zerjal et al. (2003) identified a Y-chromosomal lineage present in about 8% of the men in a large region of Asia (about 0.5% of the world total). The paper suggests that the pattern of variation within the lineage is consistent with a hypothesis that it originated in Mongolia about 1,000 years ago, and thus several generations prior to the birth of Genghis. Such a spread would be too rapid to have occurred by genetic drift, and must therefore be the result of selection. The authors propose that the lineage is carried by likely male-line descendants of Genghis Khan and his close male relatives, and that it has spread through social selection due to the power that Genghis Khan and his direct descendants held and a society which allowed one man to have many children through having multiple wives and widespread rape in conquered cities.

8% of Asian men have Genghis Khan's genes in them–It's good to be the king!

Other evidence can be found in the number of offspring from the Pharoah Ramses II who was said to have died at the age of 90 and had left behind 156 children amongst his multitude of wives. The <u>Ancient Egyptian</u> Pharaoh Ramesses II had a large number of children: 96 sons and 60 daughters–whom he had depicted on several monuments.

Now let's move onto some other animals:

Are there any other animals where females undergo menopause and live significantly longer than the age when their ability to reproduce is destroyed?

In a debate with a Selfish Gene-ist self-promoting "scientist" whose name you might recognize if I were to mention it, he made it clear he believed that menopause only occurred in humans. It did not take me long to find another example. Where do you think I looked? I looked at the other animals who, like humans, had life spans that were dramatically longer than predicted by their body size. An ugly animal it turned out to be: the bat. In addition to this I discovered that bats, like humans, also menstruate on a periodic basis! Two other species of mammals where the female has a long post reproductive life are found in pilot and killer whales. Female killer whales reach menopause before 50, but can live for up to 90 years and female pilot whales, which stop breeding by age 36, can live to be 65. I am assuming that successful male killer and pilot whales have large harems into old ages!

The prevailing mainstream dogma is that all animals experience declining fertility as they age for whatever reason. Existing theories assume that mortality increases with age, while fertility decreases in a fixed pattern:

Let's look at some other animals that do not fit into this paradigm. There are quite a few of them, consider the case of the alpine swift, a bird, that experiences increasing fertility almost throughout its life. Also, some other species grow more fertile as they age such as the freshwater crocodile and various tortoises, and several plant species, such as the agave.

Note the excellent defenses that the alpine swift has to predators:

(from Wikipedia)

The **Alpine swift** (*Tachymarptis melba*) or *Apus melba*, is a species of swift. Alpine swifts breed in mountains from southern Europe to the Himalaya. Like common swifts, they are strongly migratory, and winter much further south in southern Africa. The Alpine Swift happens to be the fastest of the Swift family.

Swifts have very short legs which are used for clinging to vertical surfaces. They never settle voluntarily on the ground, spending most of their lives in the air living on the insects they catch in their beaks. Alpine swift are able to stay aloft in the air for up to seven months at a time, even drinking water "on the wing". Their vital physiological processes, including sleep, can be performed while in continuous flight.

Let us ponder why some species might have increasing fertility with age and an increasing chance of surviving as they get older.

I believe that there is a general rule of evolution that goes like this:

-If a species evolves a defense to predation that is very effective over long periods of time, it will first evolve longer female lifespans past the animal equivalent of the age of menopause. The lifespan of the species will increase over evolutionary time periods. Eventually, if the defense to predation is successful for a long enough period of time, eventually menopause will be selected against, and the females of the species will evolve increasing fertility with age as opposed to declining fertility.

Why should this be so? It is the evolving predator that preys upon a species that causes evolution to favor a more diverse gene pool. Diversity is valuable as a defense against the whole group going extinct as it increases the odds that an effective predator defense can evolve in at least a few individuals who can survive to reproduce another day. In groups where everyone is the same, the evolving predator can readily kill them all and cause their extinction at the local level.

Once the evolving predator is permanently defeated with an excellent defense such as full body armor in the tortoise or non-stop flight and living in extreme climates , like the alpine swift, diversity is NO LONGER an advantage. What is more important under the threat of UNEVOLVING forces of mortality, is for evolution to fine tune and tweak the one perfect defense to famine or drought, etc. Once that most perfect defense to famine or drought is found, any move away from this phenotype is detrimental. So what does evolution want in these cases where only unevolving forces of death are encountered? It wants the oldest most proven survivor to reproduce more than the younger untested individuals in a group. Thus, if the perfect defense to evolving predation has been in place for a long, long, time, you will see increasing fertility with age as we see in the tortoise, the alpine swift, the crocodile, etc.

What does this tell us about humans and bats? That our perfect defenses to evolving predation which are high intelligence and flight with cave dwelling have not been completely effective for that long of an evolutionary time. We are in the process of evolving increasing fertility with age, but it hasn't happened yet. Both of our species are waiting for that mutant female to emerge that will have increasing litter sizes after the age when menopause kicks in for most females of the species.

Apparently the alpine swift, the tortoise, and the freshwater alligator have been relatively unmolested by predators for far longer than we humans and bats.

Now let's shift focus and take a closer look at what happens during menopause in women from a hormonal perspective:

-Progesterone begins a rapid decline that ends in a crash dropping from 12 ng/ml on average at age 44 to as low as 1 to 0 by age 52. This is the major signal to start the menopause. The dramatic drop in progesterone is responsible for many of the negative effects of menopause such as night sweats, hot flashes, bone loss, sleep disturbances and weight gain. A number of doctors prescribe a 300mg pill per night of progesterone (brand name Prometrium) to combat the negative aspects of menopause. The decline in progesterone by itself is not the primary culprit of the menopausal problems but rather the estrogen (estradiol) levels that remain elevated for a significant period after the progesterone crash. This is called unopposed estrogen. When estrogen is unopposed it can have some nasty effects as noted above.

MENOPAUSAL HORMONE CHANGES-FEMALE

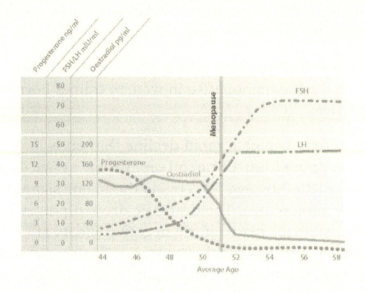

MONTHLY HORMONE CHANGES-FEMALE

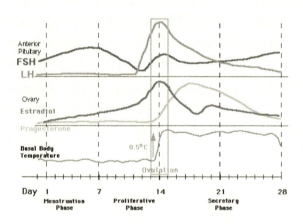

Now let's take a look at an excerpt from my paper published in 2000 in Medical Hypotheses about how we might view monthly hormone cycles as they relate to aging:

MENSTRUAL HORMONE CYCLES: A MICROCOSM OF AGING?

Early in the menstrual cycle, estradiol, while progesterone is at its nadir, stimulates the release of pituitary LHRH (a/k/a/ GnRH). LHRH first causes FSH and then LH (a/k/a gonadotropins) levels to increase. FSH alone can promote ovarian follicular development; LH is not required **(23)**. The follicle secretes estradiol which in turn triggers a rise in LH. LH causes the follicular tissue to develop into a corpus luteum which begins to secrete progesterone immediately before ovulation. Also, estradiol, FSH and LH levels crash dramatically immediately before ovulation occurs. (See Figure 1).

These events are synchronized such that LH and FSH almost simultaneously peak and then crash. Follicular rupture occurs shortly thereafter. Recently the LH surge, has paradoxically been suggested to be concurrent with a simultaneous dip in LHRH levels **(24)**. We will explore this paradox later.

The LH surge immediately precedes, and is thought to induce through luteinization, the rupturing of the follicle and the release of the ovum **(25)**. This is concurrent with a large decline in FSH. (Also note: hCG cycles with, and in the same manner as, LH and also induces ovulation **(26)**.) It is interesting to note a study that seems to suggest the logic of how LH affects the developing follicle. The study noted that LH tended to induce cAMP responses primarily in follicular granulosa cells of larger size but not in cells from smaller follicles. It also noted that apoptosis was occurring during this process **(27)**. Further, studies show that FSH suppresses follicular apoptosis in early stage follicles in a dose dependent manner. **(28)** And, finally, another study showed that hCG, like LH, can trigger apoptosis in granulosa cells presumed by the author to be initiated by inducing increased cAMP levels). (29).

The key point of these studies is that LH (and hCG as well) can drive the apoptosis of follicular tissue that eventually leads to the rupture of the follicle. If we extend this logic to the body, it can be proposed that, LH, at the high levels reached after age 40, probably drives Aging System#5 by promoting somatic apoptosis which leads to somatic atrophy and, at times, cancer. hCG likely has a similar role in promoting aging.

Now, to understand how off track mainstream aging theory is today, you have to consider that modern day evolutionary biologists, aging theorists, and the like, almost completely ignore the menopause! They have written almost nothing about it. At most they say that humans are an exception and then they don't have to discuss it! They don't even try to interpret the dramatic hormonal changes going on that drive the process of menopause. They assume it is all just an accident of evolution caused by living longer than we were designed for. And for those brave few that try to tackle menopause within the selfish gene framework, all they can come up with is the untenable grandmother hypothesis. And that's it! That's all they've got for something that so dramatically affects every human female on the planet and just screams PROGRAMMED AGING!! Some day in the future, if not immediately after the release of this book, this ignoring of menopause will be seen as a shocking failure of science. It might end up with a bunch of the other whoppers of the ages like the flat earth and the earth-centered universe!

So they have conveniently ignored female menopause and the hormone changes involved and their interpretation. This is bad enough. But let's look to the next aspect of human aging that no one has even peaked at except me as far as I know: hormone changes that occur in human males at about the same age that females undergo menopause.

What a coincidence! You see almost the same pattern of hormone changes in men around age 50 and older as you see in menopausal women! The major exception is that estrogen keeps rising in men as they age, and their progesterone takes a little longer to crash. But as far as the major sex related hormones FSH and LH are concerned, they do almost the same thing in men as they do in women-increase by 100's of percent.

Almost nobody is thinking about this. I had to search high and low to find just a single study that shows what happens to progesterone levels in men based on their age! And no charts have ever been made of it so I had to make my own charts form the data below. I fashioned the charts that soon follow from the Moroz study of sex hormones in aging males. I did find this single chart about how most male sex hormones vary with age adapted by Dr. Ward Dean. For some reason he left out the progesterone data. But this was about the only thing I could find on the internet describing male hormone changes in their equivalent of the female menopausal years: ages 50 on up.

But what is amazing is that men get almost the same changes as women , yet they have no uterus or ovaries to attack! What is going on here? No one in mainstream science seems to care or even wonder about it. But from our new perspective, the same hormone changes that are causing menopause to occur in the female, and attacking her body as a side effect, are also attacking the man's body, seemingly as a side effect. But a side effect of what? Men do not go through an abrupt reproductive cessation.

In women, after her reproductive ability has been eliminated by menopause, whatever happens to her after that time has no effect on evolution. Why? Because she can no longer pass on any of her post-menopausal traits to her offspring, as she has no more offspring after menopause by definition. Evolution does not care one iota about her anymore! Sad but true. So it might be thought that evolution just forgets to turn off the hormones that destroyed her reproductive capacity because it no longer knows she exists! So we might assume that the continued high levels of FSH and LH that then go on to attack her body are just accidentally left "on" and inadvertently kill her in due time. If this were the case, we could then say maybe the selfish gene theorists are right. That evolution does not kill on purpose, but just by accident.

But if we now look at the aging male and see that he gets the same deadly hormone changes that cause female menopause, we then just have to stop and ponder this perfect clue that evolution has given us. If aging and death were just an accidental side effect of the menopause in females, then we should not see the same deadly hormone patterns occurring in men. The existence of these hormone patterns in men, who do not have a menopause and can reproduce into old age, screams that aging and death is programmed for a purpose. In the male's case, it can still be to prevent him from reproducing too much and hurting the diversity of the gene pool, but in his case, aging and death can be considered not an accidental side effect of stopping his reproduction, but the actual tool evolution uses to stop his reproduction.

Why are males given the power to potentially reproduce many, many, times more than any human female could? It all boils down to the idea that the purpose of males is a defense to evolving predation. If there is a major onslaught by predators and most males are killed, you will need the few superior survivors to pass on their genes to as many females as possible as quickly as possible to prevent extinction of the group.

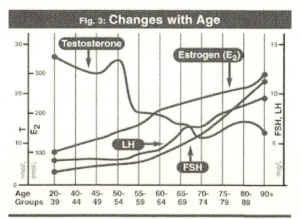

Age-related alterations of hormonal profiles in males with age. Note that the progressive rise in FSH and LH begins prior to a drop in testosterone. This is believed to be due to the progressive loss of hypothalamic sensitivity to feedback inhibition by testosterone (Adapted from Dilman and Dean, 1992, based on data from Moroz and Verkhratsky, 1985).

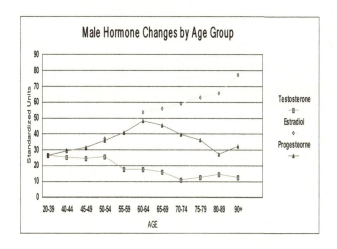

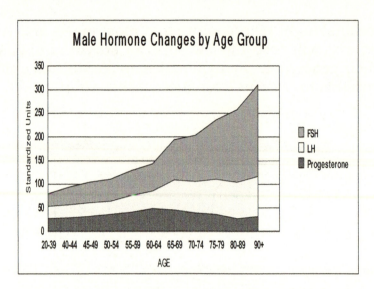
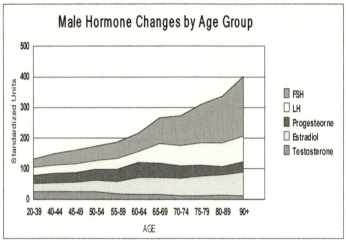

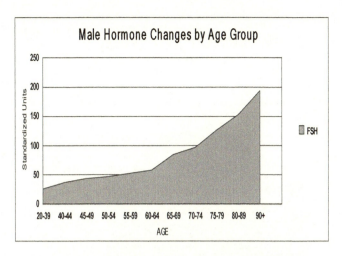

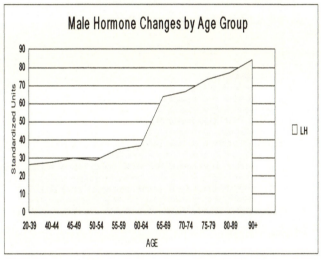

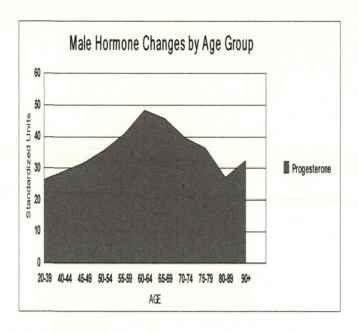

More on Progesterone-

When organizing your thoughts into a book, you have to go over old information and rethink it and then put it down in words. And from time to time, something new makes itself known to you that you did not notice before. I am coming to the realization that progesterone is likely a major anti-aging hormone whose purpose is to offset the detrimental effects of various reproductive/aging hormones. The last chart on the prior page provides a fantastic clue, where it shows men's progesterone levels continue to rise on average until about age 65, and then start to decline. It is right at this age that male LH and FSH really make a big jump up. I believe that the progesterone increase through age 65 in males is what has allowed them to live past age 50. This is likely the hormone change responsible for enabling the "son-king" to father so many children at ages where it is impossible for a woman.

Remember, a woman's progesterone level crashes to almost zero during menopause.

What one might be wondering is if LH and FSH and hCG are so bad for you and drive the aging program, how can humans, and females especially, survive such high levels of these hormones during her lifetime prior to menopause? I earlier found some studies that showed when the pituitary gland makes LH and FSH in the absence of testosterone or estrogen that the LH and FSH molecules become larger, more bioactive, and stay in the blood much longer, than FSH and LH made in the presence of youthful levels of testosterone or estrogen. So basically as you get older, your LH and FSH become much more toxic! I never did see any studies that examined progesterone for this effect, but I predict that progesterone will be shown to be even more effective at "de-toxifying" LH and FSH-the studies just need to be done.

In fact, progesterone, as mentioned before , seems to "detoxify" estrogen during menopause and can stop many of the deleterious effects of unopposed estrogen including hot flashes, night sweats, osteoporsis, weight gain, etc.

Progesterone also "detoxifies" testosterone by preventing its conversion into the "bad" form of testosterone called DHT (di hydrotestosterone). DHT is responsible for male pattern baldness, acne, and even prostate cancer. Melatonin supplements in males, lead to an increase in progesterone production which then causes reduced DHT levels. This in turn will cause a balding man to regrow a nice head of thick hair. Melatonin is known to suppress FSH and LH levels as well. Big Pharma noticed this effect of progesterone on men, and instead of telling men with prostate issues to take progesterone, they simply tweaked the progesterone molecule a tiny bit so they could patent two expensive drugs, proscar and avodart. They also noticed that men taking Proscar regrew their hair so they then started marketing it for hair loss under the brand name Propecia. Here are the chemical structures of progesterone vs Proscar and Avodart:

Proscar

Avodart

So what Big Pharma did was to take progesterone and modified it just enough to be patentable by adding a nitrogen in the bottom ring, and to clip off the CH3 at the top and replace it with something a little different. In all other respects the moclecules are the same. (The progesterone also has the little H's but sometimes in organic chemical structures they are omitted for brevity).

Unfortunately it turns out that Proscar and Avodart have some nasty little side effects. In men who take the drugs to treat prostate enlargement (BPH), or to prevent prostate cancer, if they ever do end up getting prostate cancer it tends to be of a much higher grade and much more serious than if they had taken nothing at all! It is quite possible that prescribing progesterone instead would eliminate these lethal outcomes, but hey, there is no money in progesterone since it cannot be patented! (Non-profit studies are needed on this please!).

And finally, it turns out that progesterone is likely an effective treatment for both Alzheimer's disease and ALS. I have written a book about treatments for Alzheimer's disease that describe how suppressing LH with Lupron shots prevented women with Alzheimer's disease from progressing over a 6 month period. The book also describes the case of two Brazilian identical twins who both got Alzheimer's disease at the same age. One twin took 6mg of melatonin a night to sleep, the other didn't. In a few years, the melatonin taking twin did not progress, while the twin who did not take melatonin had a devastating decline. Taking melatonin causes increases in progesterone production, and this might be the main way that melatonin performs its anti-aging miracles. And finally while I was looking for a treatment for ALS by reading all the abstracts about ALS in the Pub Med database, I noticed an interesting fact. It turns out men get ALS in a 4 to 1 ratio as compared to women, except after the age of menopause, the ratio becomes 1 to 1! This suggested to me that high lifetime progesterone levels protected women from ALS until after menopause when their progesterone crashed to about 0. A nice hypothesis with no proof at the time. Then about a year after I had put out a book on the idea some Korean scientists found that giving progesterone to male mice with a form of ALS increased their survival rate from the human equivalent of 2 - 4 years to 17 years! The right dose for them was 4 mg per kg, higher and lower levels of progesterone administration did not extend lifespan-it was a narrow window for some reason. In humans, it is likely a dose of much less than 4mg/kg will be correct as mice can handle much higher levels of drugs-more research on this is needed.

Maybe some day we should rename progesterone to protectorone to more accurately describe its functions.

CHAPTER 8: Puzzle Piece #8-The Existence of Aging Genes

The first seven puzzle pieces of the BIG PICTURE that have just been presented to you will not really raise any cackles or howls of protest from the mainstream evolutionary biologists, or Dawkins. They will do what they always do, just ignore them! They generally ignore homosexuality, asexuality, changes in hormones with age, you name it: out of sight, out of mind.

Or in rare cases they might spin some cockamamie tortured-logic explanation that often requires quite a bit of suspension of disbelief to swallow, like the grandmother hypothesis, to explain that major glaring flaw in the selfish gene argument - menopause. Or they might hang their hat on the existence of sex by saying there might be an even more selfish gene than all the others-the sex gene! (That reduces the spread OF ALL YOUR OTHER GENES BY at least 50% (at best-assuming you had an immediate unlimited supply of mating partners at all times).

Or they will try to explain the existence of females' preference for bright colored males by saying the bright colors reflects a healthy immune system! A pretty far stretch to anyone outside the evolutionary biology community.

But what does this do for them? It allows them to keep plugging away at their little selfish gene math models with all sorts of complicated equations and statistics and to wallow around aimlessly in their myopic, complex, microscopic world of the selfish gene.

While the first seven puzzle pieces of the BIG PICTURE previously described will just elicit yawns or glassy stares from the mainstream theorists, this eighth puzzle piece, the existence of aging genes, will swiftly provoke howls of ridicule and derision from them. It addresses the one thing they think they are experts at-the gene. They will not ignore this eighth puzzle piece of the existence of aging genes because it contradicts everything they think they know about how evolution works. In fact I was told in an email once by Brian Charlesworth a well-known biologist, that "you do not know how evolution works" when I tried to explain to him how a paper by Sir Peter Medawar on aging was missing half the picture. And he was right! -lucky for me.

You see when one's whole career is spent proving different ways that evolution works by selfish genes trying to make more copies of themselves, the idea of a gene that exists with the sole purpose of killing you and exists for the sole purpose of restraining the spread of your genes at some time in your life just does not compute! The obvious reason is that it will reduce the spread of your selfish genes, and selfish genes never do that! In their minds selfish genes only want to increase their number as much as possible. So to the mainstream selfish gene theorists, aging genes and thus an aging program are impossible to evolve and thus cannot exist. To them aging can only be an accident of evolution of animals living longer than they do in their normal (wild) environment.

AGING GENES?!@#!! ??

THAT DOES NOT COMPUTE!!!@#@!!

So this is the big impasse that I have reached by working the theory of evolution from a top down approach, from the BIG PICTURE down to the little picture of the gene. This is the big impasse that the mainstream evolutionary biologists have reached from working the theory of evolution from the bottom up approach, from their trying to start with a little single gene and working it up into bigger and bigger levels of complexity to explain everything about the BIG PICTURE-it cannot be done.

They are sure that aging genes cannot exist. It contradicts everything they think they know.

I am sure that aging genes must exist or else the entire BIG PICTURE that has been created so far, is incorrect. And everything I know is wrong.

So let's look at the evidence.

The funny thing is that these selfish gene theorists have recently been producing increasing amounts of evidence that aging genes DO exist, yet they ignore them when the facts stare them right in the face.

For example, a recent study was just published in October 2015, with the lead researcher being Brian Kennedy of the Buck Institute for Aging which showed that in yeast, there are 238 genes that WHEN REMOVED **INCREASED** the lifespan of yeast, and that many of these genes were also found in humans and never before known to be associated with aging.

And what did they name these genes? LIFESPAN GENES!! When in fact we can see quite clearly that they are AGING GENES! This is how blind the mainstream theorists are to the truth. Even the title of the following article is wrong, it should be "Scientists Identify 238 genes linked to DECREASED lifespan". Why? Because the genes they found only increase lifespan when they are removed! Their self-deceptions really stupefy me to no end.

From the website medicaldaily.com :

Vitality

Aging Research Update: Scientists Identify 238 Genes Linked To Increased Lifespan

Though we are well-acquainted with the aging process of our own bodies, only the scientists among us ask: What is happening on the cellular level? A <u>new study</u> has identified 238 genes that, when removed, increased the lifespan of yeast cells (specifically their ability to continue replicating themselves). While yeast cells may be a far cry from human cells, many of the cellular mechanisms found in their genes continue on, surprisingly enough, through higher organisms including human beings.
Importantly, 189 of these genes have never before been linked to aging. Someday, then, this valuable research may be applied to human health.

Lifespan Genes

The study began with an examination of 4,698 separate yeast strains, each with a single gene deleted. The researchers designed the study to answer a simple question: Which of the strains increased lifespan? And so they observed and recorded the entire life cycle of each microscopic strain. Every time the mother divided, the researchers used a needle (attached to a microscope) and teased out the daughter cells, and then counted how many times each of these new cells divided.

This detailed work, funded by the National Institutes of Health, was performed by the laboratories of Dr. Brian

Kennedy, lead author and the Buck Institute's president and CEO, and Dr. Matt Kaeberlein, a professor in the department of pathology at the University of Washington.

Their painstaking efforts showed how different genes (or the absence of them) modulated the aging of yeast. For example, one gene called LOS1 appeared to have a particularly strong impact on lifespan.

LOS1 is known to be modulated by mTOR, a growth regulator inside each cell that, based on nutritional and environmental cues, helps to orchestrate energy levels, stress, growth, and amino acids — key factors associated with a cell's calorie use and lifespan. In turn, LOS1 influences a gene, Gcn4, that helps modulate DNA damage control. LOS1 also helps relocate transfer RNA, which performs the job of transporting amino acids to ribosomes in order to help build proteins, the worker bees of a cell.

Because a number of the genes identified by Kennedy and Kaeberlein are found in C. elegans round worms, this suggests some of the same mechanisms may exist in higher organisms, possibly even humans.
"Many of the longevity genes identified here are known to act in longevity pathways conserved in multiple species," wrote the authors. "...Others, like the LOS1 pathway, are largely unstudied." (end article)

There have been quite a few other aging genes discovered that when knocked out significantly increases the life span of the carrier such as the p66shc gene in mice when knocked out, lowers the level of oxidative stress in the mice and increases life span.

There is a researcher by the name of Cynthia Kenyon whose laboratory is cranking out all sorts of evidence that aging genes exist. She has mutated a single gene (daf-2) in the round worm C. Elegans and that mutation caused its life span to double! She also found a gene that when mutated in dogs limits their size (the mutation creates all the small dog breeds), she found it also controls aging and it explains why smaller dogs have a much longer life span (25 years +) as compared to bigger dogs (10 years+).

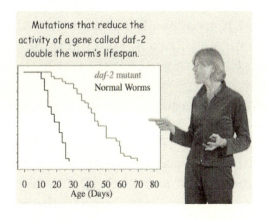

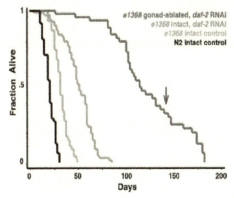

Fig. 2.16. Worms that live six times as long as normal. Arrow, two worms day 144. Redrawn from Arantes-Oliveira et al. (2003).

Other researchers have discovered genes that when knocked out increase the life span of the organism by a factor of 10! (see Dorman J, Albinder B, Shroyer T, Kenyon C. The age-1 and daf-2 genes function in a common pathway to control the lifespan of Caenorhabditis elegans. Genetics 141(4), 1399-1406 1995 AND Ayyadevara S, et al. Remarkable longevity and stress resistance of nematode PI3K-null mutants. Aging Cell 1:13-22 2008).

The writing is on the wall for all to see ...Aging genes do exist. How do the selfish gene theorists try to explain this away? They say that even if there is a gene that you can delete that increases life span it "must" reduce the number of offspring (fitness) somehow that the animal might have if it lived in wild conditions!

They are using a variation of an old argument invented by the biologist George Williams in 1957 called antagonistic pleiotropy. Pleiotropy just means a single gene has multiple unrelated effects on an organism, and antagonistic means some of the effects are bad. In reference to his hypothesis he applied these two concepts to single genes, where a gene was suggested to have a good effect on young individual (it allowed the young individual to increase the number of offspring he or she could have). At older ages the gene had detrimental effects (aging) on the individual by reducing the number of offspring he or she could have. Given that he had few facts to work with back then, it seemed a reasonable hypothesis, but since then there have been no genes yet discovered that actually have this Dr. Jekyll and Mr. Hyde personality with respect to aging vs. fertility.

(In fact, it appears Williams was dimly noticing that there was some sort of link between reproduction and aging when devising his theory and blaming the link on genes that have two functions to-

1. promote reproduction and then to

2. accidentally cause aging at a later date.

Maybe he had briefly pondered the fact of menopause before coming up with his hypothesis. What I believe he was trying to describe and what I have shown earlier, is that the hormones that drive reproduction also drive the aging process at a later date. So his mistake was to blame the antagonistic pleiotropy on a single gene when it really is being driven by the reproductive/development hormones - good for you and reproduction at younger ages, deadly at older ages.)

There was an excellent study done by Marc Tatar with fruit fly larvae. Tatar figured out a way to deactivate the receptor for a development/reproduction related hormone called Juvenile Hormone (JH). Like LH in humans, JH drives fruit fly larvae through their fly equivalent of puberty. He applied JH to the larvae with the inactivated receptors for a while, as expected no development took place. Later he reactivated the receptors and the JH caused the fruit flies to develop into adults. BUT IT ALSO TURNED OUT THAT BEING EXPOSED TO JH CAUSSED THE LARVAE TO AGE AS WELL so the adult flies had much shorter lifespans. This study basically proves the concept that reproduction related hormones have a second function in driving the process of aging. If there ever is a Nobel Prize for aging studies, this study would deserve the first nomination.

Also as an aside, George Williams single-handedly set back the progress of evolutionary theory at least 50 years when he published a scathing critique about the impossibility of group selection as a force of evolution in 1966 titled Adaptation and Natural Selection-so hostile that subsequent young evolutionary theorists dared not even mention the idea or they might see their careers in ruins-this will be discussed later.

George C. Williams

Single-Handedly Set Back Evolutionary Theory by 50 Years!

When confronted with the existence of aging genes, the selfish gene theorists will not really address the evidence directly but rather invite you into a game of mental masturbation by challenging you to explain how an aging gene could possibly evolve and not be immediately selected against by evolution.

How an aging gene could evolve is actually a good point and initially a mystery and does require a bit of thought to figure it out. Actually Peter Medawar an early aging theorist pioneer had it right when he suggested mutations in genes that are expressed at old ages could be the source of aging.

Sir Peter Medawar
Nobel Prize for Medicine 1960

"It is naïve to suppose that the acceptance of evolution theory depends on the evidence of a number of so-called "proofs"; it depends rather upon the fact that evolutionary theory permeates and supports every branch of biological science, as much as the notion of the roundness of the earth underlies all geodesy and all cosmological theories upon which the shape of the earth has bearing. Thus, anti-evolution is of the same stature as flat-earthism."

I kind of feel bad that I wrote a paper about his seminal aging paper. My paper was titled 'Shattered: Medawar's Test Tubes ad Their Enduring legacy of Chaos". It looked like I was picking on him. But of all the early aging theorists he was actually probably the smartest, and did win a Nobel Prize for pioneering human organ transplant technology. Sorry Sir Peter, nothing personal.

It is much easier to understand what Medawar was suggesting with a real world example.

There is a disease called Huntington's disease that has an average age of onset at about 40 years old. It is fatal after about 20 years, it is caused by a mutation in the Huntington gene and can be passed onto children. It is neurodegenerative and it causes its victims to have all sorts of mental issues as well as causing their bodies to jerk around in a fashion that is described as similar to dancing. Thus Huntington's disease used to be called Huntington's Chorea. (Chorea as in dance/choreography/etc.) While the average rate in the population of people getting Huntington's disease is 7 per 100,000 there are pockets of people around the world where the incidence is much higher. For example, one of the highest incidences is in the isolated populations of the Lake Maracaibo region of Venezuela, where HD affects up to 700 per 100,000 persons.

Research In Venezuela

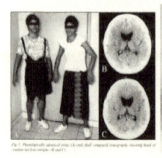

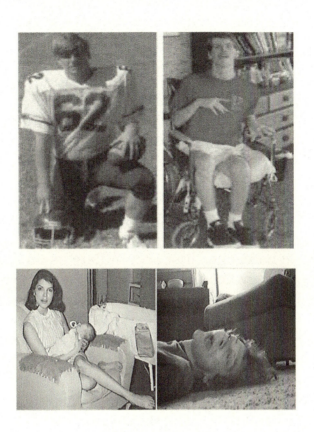

Before And After

This high incidence of Huntington's Disease in parts of Venezuela, affecting almost 1% of the population means people with the gene are reproducing quite a bit and it cannot be blamed on just a lot of people having new random mutations occur in their Huntington's gene. The gene is getting passed on to offspring at a high rate. From this we can see how aging genes might evolve and spread in the population. If they are detrimental to you at a much later age than when you first start reproducing, they can accumulate in the population with little to no effect on a person's evolutionary fitness (evolutionary fitness just means how many kids you have).

So if you can imagine a whole bunch of these late-acting killer genes building up in a population's gene pool that only activate after the age when most individuals stop reproducing , then you can see how bad genes can build up in a species' DNA and be passed on to offspring.

So "no big deal" at this point, the mainstream selfish gene theorists would say. They would suggest that there is no overriding program designed to kill you here, but rather these bad genes just accumulate as an accident of evolution. They accumulate because most people don't live long enough to reproduce after the age that these genes get switched on, (so there is nothing selecting against those with the bad genes or FOR those without the bad genes-if you are not reproducing, evolution does not care about you) .And they go on to say that the few people who do live long enough to reproduce at old ages they suggest , are too few in number to affect the gene pool one way or the other. (Of course they ignore menopause here in humans so we know this would only apply to men).

So at this point I don't really have a problem with mainstream theory, I too believe that aging genes initially are not selected for by evolution, but arise randomly by accidental mutations.

Now here is where I will part company with the selfish gene theorists and I will affirm that even though aging genes arise as random mistakes of evolution, these detrimental aging genes are eventually CAPTURED by evolution and made part of a CONSERVED program of aging that is retained for the benefit of the group at the expense of the individual-THERE I SAID IT! And I'm not taking it back!

Selfish Gene Theorists cannot swallow this idea either, it is called group selection and it is wildly unpopular amongst them. Anyone promoting the idea of group selection being something other than just a minor side show in evolution is guaranteed to have a quick ride out the university exit door to a new career.

What is group selection? It is a form of evolution that goes on at the group level where say one group of rabbits who ages, somehow outcompetes a group of non-aging rabbits. So instead of selfish genes guiding the rabbits' evolution under the force of group selection apparently the genes (aka the genome) are no longer selfish and are willing to sacrifice themselves for the good of the group. Very unpopular amongst the selfish gene theorists as you could imagine-but it is the underlying idea that leads to the title of this book...

THE UNSELFISH GENEOME

The rest of this book will be about how it is possible for an unselfish genome to exist, how it works at the DNA level, how it could have been selected for by correcting selfish gene theorists' incorrect version of what group selection is. Basically I will be working backwards from the BIG PICTURE and forcing what we know about genetics and DNA, gene regulation, and timing of gene expression to fit into the reality that there is an aging program created randomly but captured by evolution that **is designed to kill you at a certain age**, or at least to stop you from contributing too much to the gene pool. And I will then do the impossible.-Show how sex and aging, although a subtle force, is selected for almost everywhere and almost all the time by evolving predation.

In other words, the rest of this book will be concerned with finding out what is wrong with the selfish gene theorists' little picture that keeps it from being ramped up to explain the BIG PICTURE that sex and aging evolved as defenses to evolving predation.

And before we move onto the next chapter , let me just share with you some evidence that the newer, younger researchers into aging are finding that the evidence is indeed building up against the selfish gene theorists (don't they sound malicious?? "The selfish gene theorists!"

An excerpt from a 2009 article posted at the Dailygalaxy.com

"Prevailing theory of aging challenged by Stanford University Medical School researchers.

Their discovery contradicts the prevailing theory that aging is a buildup of tissue damage similar to rust.

The Stanford findings suggest specific genetic instructions drive the process.

If they are right, science might one day find ways of switching the signals off and halting or even reversing aging.

"We were really surprised," said Stuart Kim, who is the senior author of the research.

Kim's lab examined the regulation of aging in C. elegans, a millimeter-long nematode worm whose simple body and small number of genes make it a useful tool for biologists. The worms age rapidly: their maximum life span is about two weeks.

Comparing young worms to old worms, Kim's team discovered age-related shifts in levels of three transcription factors, the molecular switches that turn genes on and off. These shifts trigger genetic pathways that transform young worms into social security candidates.

The question of what causes aging has spawned competing schools, with one side claiming that inborn genetic programs make organisms grow old. This theory has had trouble gaining traction because it implies that aging evolved, that natural selection pushed older organisms down a path of deterioration. However, natural selection works by favoring genes that help organisms produce lots of offspring. After reproduction ends, genes are beyond natural selection's reach, so scientists argued that aging couldn't be genetically programmed.

The alternate, competing theory holds that aging is an inevitable consequence of accumulated wear and tear: toxins, free-radical molecules, DNA-damaging radiation, disease and stress ravage the body to the point it can't rebound. So far, this theory has dominated aging research.

But the Stanford team's findings told a different story. "Our data just didn't fit the current model of damage accumulation, and so we had to consider the alternative model of developmental drift," Kim said."

Finally, before we leave this topic for the next chapter, I think it might be a good idea to just run through a simple thought experiment to understand how aging is popping up in all species of animals all the time with no guidance from evolution. It is called mutation. When a gene mutates it is caused by a random mistake in its sequence, often a copying error. For example if there was a sentence on a page (let's say it represents a gene), and it reads:

"Start-Grow lips on the fetuses face-End"

and let's say it was accidentally copied as

"Start-Grow hips on the fetuses face-End"

(this would be called an l to h mutation)

You can see how a minor mutation can end up being disastrous. And you can also see how most mutations would have a negative effect or at best neutral. **The positive mutation is a rare occurrence**. So evolution has to go through a lot of bad mutations to find a good one. When bad mutations occur, they are often fatal, and at best debilitating, and in the wild, predators make sure they are quickly disposed of.

Okay, so now let us begin our thought experiment with a hypothetical animal that experiences a long sequence of genes that are activated. There have to be genes that are only activated well after the animal is able to reproduce. In a human, for example, this might be a gene for beard growth, or baldness, or hair graying, nose or ear growth etc. In other animals these could be genes that are involved in some sort of metamorphosis.

Anyway, let us now imagine a group of these animals living peacefully in a predator-free environment. They live and reproduce freely and only die eventually from an accident or natural disaster-they do not age.

Now, in the next chapter we will learn how gene expression can be controlled by adding methyl groups to them so that the genes are activated only when the methyl group comes off. It will be explained that this is how the genes of these animals are all controlled to fire in a sequence over a life time. Just take this as a given for now and let us move on.

Now along comes a predator that we will say can catch and kill so many of the animals that the average animal will only live to say 40% of its expected life span in the predator free environment. If these were humans you would say they are getting killed off usually by age 40 instead of living to say 100 (let us say age 40 and age 100 instead of using percentages from now on for ease of description).

Now if our animals can start reproducing at say age 20, the population can still survive even though everyone's life has been cut short at age 40 on average by predation.

Now what is happening to all the genes that were timed to be expressed after age 40? They are undergoing random mutations that are usually bad and harmful. What were good genes are just randomly turning into aging genes. The genes are still methylated so they still retain their timing mechanism, but after a bad mutation in that gene, what was once good for the animal, is now an aging gene timed to be expressed at older ages. After you remove the predator and allow the population to grow older again, the good genes, say for growing a beard, might have mutated into an aging gene that causes the hairs to grow inward and eventually pierce the animals brain or in human terms maybe it now causes Alzheimer's disease.

So that is how aging genes can form, and the selfish gene theorists do not disagree, they just call it aging due to mutation accumulation. Aging genes are being randomly generated in all animals all the time and are often selected against if the animal lives long enough, but in animals that die early, aging genes can accumulate and thus aging can evolve.

Now if the selfish gene theorists were correct, evolution would quickly select against these aging genes to allow each individual to spread copies of his/her genes as far as possible. However this is not what happens. Apparently evolution has figured out a way to capture and conserve this ACCIDENTAL creation of aging genes for the good of the group, even though it is at the expense of the individual. Proof of this idea is that genes related to aging that are similar are found in humans, bacteria, worms, yeast, you name it. It seems aging must have evolved long ago in an ancient ancestor to most life forms, and it has been captured and conserved by evolution and allowed to spread throughout the animal kingdom! And it must be really hard to get rid of! It has likely been stitched into the fabric of our entire existence. If not, aging could easily be discarded by evolution in a very short time frame. Thus, aging has been stitched by that sneaky force of evolution into the act of reproduction so that it would be very hard to lose. Like a which came first problem- the chicken or the egg? The easy way to get rid of aging would be to get rid of the hormones that activate aging genes. But since the same hormones that cause aging also drive reproduction, you have a Mexican standoff. Nothing is going to happen easily or quickly! <u>Aging is very hard to get rid of because when you do, you also reduce your reproductive output! Clever this force of evolution!</u> I won't cite it but if you want to look, evidence for this idea is everywhere!

Wait, I will give you this little example: Most bacteria, have DNA that forms a circle, called a plasmid. No beginning, no end. So when it makes a copy of itself, it just makes a new circle of DNA. Now after bacteria evolved into plants and animals, called eukaryotes, the DNA was no longer circular but instead it was linear, we call linear DNA chromosomes. There is a problem with chromosomes, they cannot be copied all the way to their end so every time they are copied they shorten. So sooner or later the DNA shortens so much that it destroys a gene and it kills the host. So you see every time a eukaryotic cell reproduces and copies its DNA it also ages by DNA shortening. <u>Evolution has embedded aging into reproduction at all levels.</u>

(Keep in mind that this example is overly simplistic when discussing modern telomeres which have added all sorts of bells and whistles, but this analysis most likely is valid for the first rudimentary telomeres that evolved when circular DNA first became linear).

CHAPTER 9: Puzzle Piece #9 Aging Can Be Slowed & Even Reversed.

If you listen to the selfish gene theorists about aging, you would think that stopping or even slowing, much less reversing, aging would be hopeless. Why? They assume that one's DNA over their lifetime undergoes damage that accumulates and it is basically irreversible. Their idea of intervening in the aging process can be likened to trying to put a firecracker back together after it has exploded. Somehow they are able to maintain this position even when faced with facts that fly in the face of this assumption.

From a programmed aging point of view, intervening in the aging process is much simpler, it only requires identification of the fuse and cutting it before the explosion starts.

One glaring fact that has been sitting right in front of the selfish gene theorists for as long as they have been theorizing is the existence of the phenomenon of caloric restriction.

Caloric Restriction:

It has been well known since the 1930's that if you underfeed a rat or a mouse, that they can live up to 40% longer than a normally fed rodent. Life span extension through underfeeding was discovered by Clive McKay in the 1920's who was originally a fish farmer who noticed that his farm-raised trout lived longer when his helper forgot to feed them while Clive was on vacation. He wasn't sure if it was the underfeeding of the fish, or the protein source they used for food which was bugs that flew into their pond that slowed their aging.

He applied a similar regimen of underfeeding to a group of young rats and found a dramatic life extension effect by underfeeding them to the point where they could not grow. He eventually would allow them to gain 10 grams of weight from time to time when they started to look unhealthy. After the experiment he found that the regimen added a large amount of extra lifespan to the males, but for some reason it did not increase the lifespan of females significantly.

Clive McCay

THE EFFECT OF RETARDED GROWTH UPON THE LENGTH OF LIFE SPAN AND UPON THE ULTIMATE BODY SIZE [1]

C. M. McCAY, MAEY P. CROWELL AND L. A. MAYNAKD

Animal Nutrition Laboratory, Cornell University, Ithaca

ONE FIGURE

(Received for publication January 18, 1935)

In a preliminary report, the literature concerning the effect of retarded growth upon the life span was reviewed (McCay and Crowell, '34). In this report was also included a sum mary in the nature of a progress report dealing with a study employing rats to determine the effect of retarding growth upon the total length of life.

The present summary represents a complete, final report of this experiment employing white rats and covering a period of nearly 4 years. The object of this study was to determine the effect of retarding growth upon the total length of life and to measure the effects of retarded growth upon the ultimate size of the animal's body. In the present study, growth was retarded by limiting the calories.

The growth of an animal can be retarded either by disease or by various nutritional deficiencies. Every laboratory that performs vitamin assays is familiar with the retarded growth and prompt death that result when there is a deficiency of a certain factor such as vitamin A. In such experiments the animal grows little and dies prematurely. However, a border

line level of such an essential as vitamin A may permit a very slow growth and permit the animal to approach or attain

[1] These studies were supported in part by the Snyder research grants and we appreciate the assistance of Mrs. Harry Snyder in making these studies possible.

(The first page of his famous paper).

FIGURE 1

Survival curves generated from the data of McCay et al. (1) for male (*A*) and female (*B*) rats consuming ad libitum (all you can eat) vs. calorically restricted

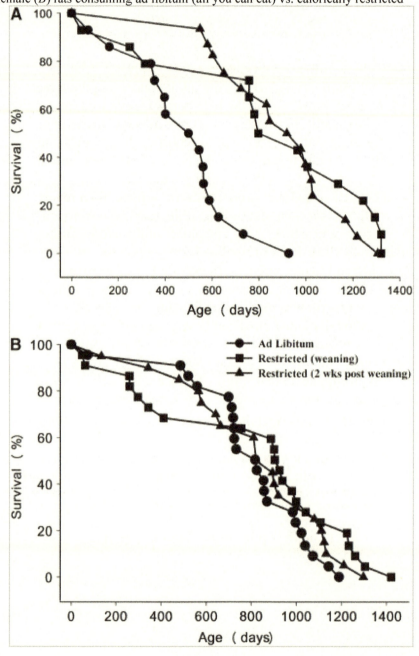

The method is quite simple, you just measure the amount of food that a normal mouse or rat eats when it is allowed to eat all it wants, and then just reduce the amount by a certain percentage. Typically the maximum reduction in food intake before it kills the animals is about 40%, which also corresponds to a 40% decrease in their body weight compared to the well fed animals.

A 40% increase in both average and maximum life span is quite a dramatic proof that aging is not a fixed process that cannot be altered! Yet the selfish gene theorists like to explain this effect by saying something like the reduced food intake produces reductions in free radicals which reduces the amount of DNA damage and thus the rate of aging is slowed!

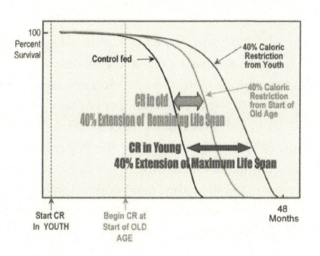

Here are two same-aged monkeys: one semi-starved for life/the other eating all it wants. In monkeys, the major hormone changes noted by researchers caused by caloric restriction was an increase in melatonin and DHEA.

However, I believe the hormone changes from caloric restriction are much more profound than just a rise in melatonin and DHEA. While researching caloric restriction for my 1998 paper, I came a cross a study where they starved a number of Navy men with a complete fast of 0 calories for 5 days. They noted the following hormone changes by day 5: (note in my 1998 paper I noted that hormones that stimulated cAMP directly, or caused a transient rise in cAMP in cells tended to be the "bad" hormones that rose with age. While the cGMP associated hormones tended to be the "good" hormones that decline with age.

cAMP stimulating hormones:

TSH declined by 67%-as expected

LH decreased by 33%-as expected

FSH decreased by 33%-as expected

hCG (not tested)

cortisol increased by 110%-unexpected

estrogen -increased by 10%-unexpected

cGMP stimulating hormones

Melatonin increased +/-100% in rats -as expected (this was not tested in the Navy study)

GH increased 200%-400% in men -as expected (this was not tested in the Navy study)

DHEA-S increased 100%-expected

Testosterone-decreased 50%- unexpected

T3 and T4 were relatively unaffected, and prolactin declined 25% but is not listed because it is an "ambidextrous" hormone stimulating both cAMP and cGMP depending on which receptors it influences.

A prediction I made in my 1998 Paper- The Evolution of Aging-etc., that I really didn't believe that much at the time, was that caloric restriction would increase life span by causing the body's levels of cyclic guanosine monophosphate (cGMP) to rise.

I came up with this prediction by noting that there were two major kinds of hormonal second messenger systems. The first second messenger system is known as cAMP for (cyclic AMP). You might recognize the hormones that stimulate cAMP with the names LH, FSH and hCG-those hormones that increase with age in order to age you and kill you. It is called second messenger system because the hormone binds to a receptor on the cell, which then activates a second message inside the cell (the cAMP). I reviewed all the cAMP stimulating hormones and found almost everyone was involved in causing some sort of cancer.

Now there is a less well known and less well-studied second messenger system called cGMP (for cyclic GMP) which doesn't seem to have a lot of hormones that directly stimulate it. It is stimulated however when exposed to nitric oxide. This didn't make much sense to me, so what I did was I looked into the Pub Med science study database and searched each hormone involved in aging against cAMP or cGMP. What I found was that all the hormones that are good for you and decline with age, are associated with higher levels of cGMP. Although I never did find out how the good hormones were causing the increase cGMP.

It was a kind of a murky prediction so I never really dwelled on it that much. But imagine my surprise when the prestigious science journal NATURE published a study stating that cGMP increases dramatically during caloric restriction!

Figure 1 - In white adipose tissue and other mouse tissues, caloric restriction (CR) induces cGMP production through an increase in eNOS levels.

From the following article

(Nitric Oxide)/cGMP link between calorie restriction and mitochondria

Leonard Guarente
Nature Chemical Biology **1**, 355 - 356 (2005)
doi:10.1038/nchembio1205-355

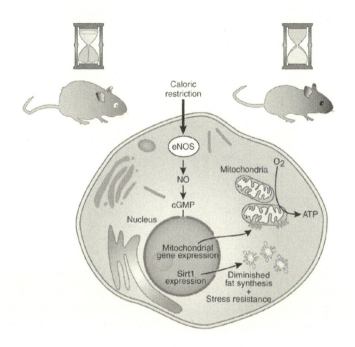

Metformin:

An age old drug that has been prescribed to diabetics for many years also seems to have life-extending properties. The generic name of the drug is metformin, also known by the brand name Glucophage. The Glucophage name hints at how it acts, it destroys or suppresses glucose. This in turn, I believe, has the effect of lowering one's blood sugar which tricks your body into thinking it is undergoing caloric restriction. So far, metformin administration has been show to extend average and maximum life span in worms and rodents. It also has the beneficial side effects of inducing weight loss, and retarding cancer in humans. In fact, diabetics who take metformin tend to survive longer than non-diabetic humans.

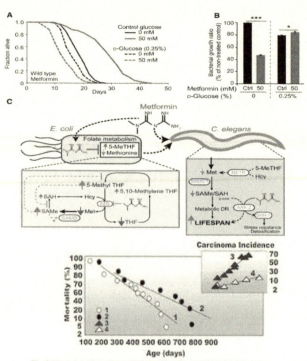

Fig. 4. Extension of maximum lifespan of and reduced tumor incidence of mice with Phenformin (a no-longer available analog of Metformin) (Dilman, 1981).

Melatonin:

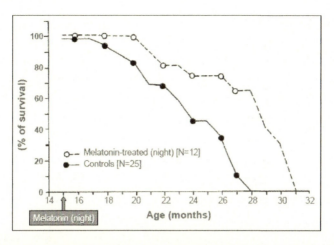

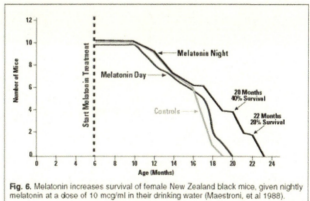

Fig. 6. Melatonin increases survival of female New Zealand black mice, given nightly melatonin at a dose of 10 mcg/ml in their drinking water (Maestroni, et al 1988).

Mice live longer if you add melatonin to their drinking water (at night only).

DHEA:

Men with higher DHEA levels have a much better survival curve see Box F below.

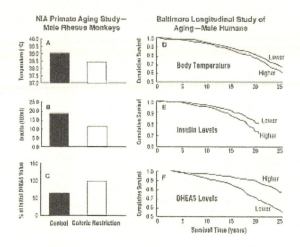

Resveratrol:

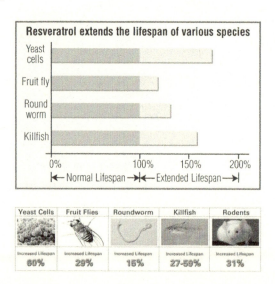

(Note: Resveratrol was only effective at extending the lifespan of mice on a high calorie diet but there was no significant life span extension for normally fed mice.)

Water Restriction:

I once wondered why evolution would cause life span extension in animals that were underfed? It made sense to me that if a group of animals was being hit by a famine that evolution would want to stop them from reproducing (the melatonin increase from caloric restriction does this nicely) so that the additional nutritional demands of having and feeding offspring during a famine does not lead to the death of all the offspring and the mother. She can wait to reproduce after the famine is over. I then thought it would also be smart if evolution would stop the aging process as much as possible so if the famine was a long one, that any survivors would still be young enough to reproduce once the famine was over. Sounds logical, right?

But then I asked myself what causes famines? And the answer popped up immediately- droughts. And droughts should be much longer than famines, as they would precede the famine and cause various drought-killed plants and animals to provide food through some of the drought before the famine started with a vengeance. Thus I predicted that water restriction should be a superior and more dramatic life span extender than caloric restriction. (I can't believe no one ever thought of this before!!)

There was almost zero research on this idea so I took it upon myself to purchase 10 female Sprague Dawley rates and conduct an experiment at home in my closet. Little did I know I was in for about 4 years of work as one of my two water-restricted rats lived to the world record age for this rat strain of 47 months! I say this is a world record because I searched all the science studies that involved Sprague Dawley rats undergoing caloric restriction, a number of them with thousands of animals, and there were no Sprague Dawley control females that lived longer than 34 months and no Sprague Dawley CR animals that exceeded 39 months. So with just 2 rats undergoing water restriction I ended up with one female that broke the record for the breed! This suggests that the possible maximum lifespan extension of water restriction might be even way more than this 47 months for the breed! I actually replicated Clive McCay's experiment by limiting water intake to control the growth of the young rats. I got a dramatic lifespan extension in females while McCay's protocol only got life span extension in males and NOT in the females! I have uploaded a YouTube video of the water restriction experiment which can found by searching for "Oldest rat in the world". I have taken a few still shots from the video and add them below:

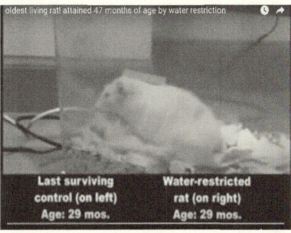

Size comparison of two wr rats to a control: age 19 mo.

Deprenyl (Seleginine):

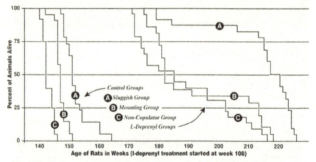

Figure 1. Survival curves of old male rats treated with deprenyl (n = 22) or saline (n = 22). The rats were grouped according to sexual activity (A—sexually active; B—able to mount, but unable to ejaculate; C—sluggish, non-copulators). Note that sexually active members of each group lived the longest; and that all members of the saline-treated group died before any of the rats treated with Deprenyl died. This study helped to earn Deprenyl the reputation of being the anti-aging aphrodisiac.

Parabiosis:

Believe it or not, it was Clive McCay who reinvestigated in 1956 another important technique that was recently resurrected for aging research in 2012 called parabiosis. It is where a young rat is joined to an old rat so that they share the same circulatory system and thus the same blood. Recent research of this seemingly cruel form of vivisection has shown that there are factors in young rats' blood that can rejuvenate the muscle of old rats. One of these substances is purported to be GDF11 (growth differentiation factor 11). Although some labs have not been able to repeat the rejuvenation effect by administering GDF11 to old rats suggesting that other substances in the blood may be involved (maybe hormones?). Other studies have recently shown that the blood of young mice can repair cognitive decline in older mice and cause old brains to rapidly generate new brain cells.

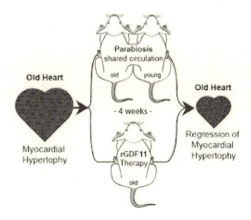

If you are interested in buying any of the life extenders mentioned in this chapter without a prescription you can get all of them from International Antiaging Systems which can be found on the internet. I have purchased from them many times and never lost a shipment. They ship from the UK to the US. You can get melatonin, resveratrol, and DHEA in the US without a prescription. And you can get very, very cheap powder resveratrol and melatonin (90-95%% cheaper than pills) from the website TakeD3.com.

CHAPTER 10: The Timing Of The Expression Of Aging Genes-via 5mC, Chromatin, Lamin A, WRN

In 1998 I published my first science journal article ever in Medical Hypotheses, one of the only science journals dedicated to theory. (Almost all other medical and science journals are dedicated to the results of experiments.) It was titled: "The Evolution of Aging: A New approach to and old Problem of Biology". In that paper I made a wild prediction that was laughed at heartily by most who first heard it: that Luteinizing Hormone (a sex related hormone) would be found to be involved in driving the progression of Alzheimer's disease. Well, it turns out that over the years that the prediction came true as explained elsewhere in this book.

Another prediction that I made in quite a bit of detail in that paper was that epigenetics would be the primary way that aging was controlled at the level of DNA. (Epigenetics is simply the control of genes by things like proteins or molecules which attach to the genes and silences them until they are removed). This was almost entirely ignored by everyone who read my paper even though it was a major theme running through its 42 pages. And guess what started happening about 15 years later? All sorts of researchers and scientists are popping up with papers describing how EPIGENETICS is controlling aging at the DNA level. See for example: <u>How does the body know how old it is? Introducing the epigenetic clock hypothesis.</u> Mitteldorf JJ.Biochemistry (Mosc). 2013 ep;78(9):1048-53.

Apparently my first paper was so far ahead of its time in 1998 that I got tired of trying to convince anybody of anything in it and took 15 years off, but I am starting to get my feet wet again now that I can start telling them "I told you so!".

So all I want to do in this chapter is to give you a very crude understanding of how I proposed that aging is controlled at the DNA level as I proposed in my 1998 paper. I will keep it simple so that even a 10 year old can understand it; I don't want your eyes to glaze over with too much detail.

So the first question we need to ponder is:

During the process of development from a fertilized single egg into a full grown human being, how are all the different genes that need to be turned on and off at the right time controlled?

There are a number of ways that this happens, but in my paper I focused primarily on a thing called DNA methylation.

What is DNA methylation and how does it control genes during development and aging?

Well let's start with a very simple model of DNA.

Inside each cell of our body, is a nucleus. If each cell was considered an individual animal (which is kind of true as single cells can be removed from the body and be kept alive in a glass dish), the nucleus would be considered the brain of the cell. And inside that brain is information. The information is encoded on a long piece of string of molecules called DNA. The DNA is really just like a book written with tiny letters on a string instead of on pieces of paper.

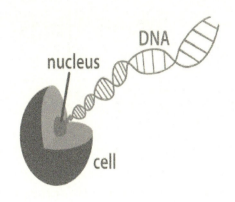

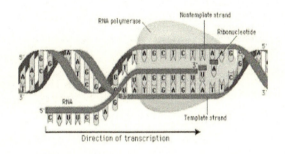

Groups of these letters make up a sentence, which is like a single instruction. Each sentence of the book on a DNA string represents a gene. So your DNA in a simple way might look like this:

GENE1----Gene2----Gene3----Gene4-----etc.

So when an egg starts dividing into 2 cells then 4 cells then 8 cells etc. all the way up until it has created a human adult, various genes need to be turned on in the right sequence. Gene1 might be something that says- single cell divides into two cells, gene2 says two cells divide into 4 cells, etc.

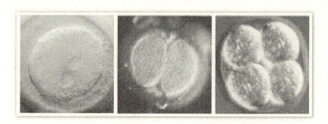

The important point here is that a huge number of genes need to be turned on at just the right time to properly have an egg go through the complicated process of developing into a full grown adult consisting on average of 37.2 trillion cells. This sounds like a lot of cells but it can be created by starting with one cell and just having it and its daughters all keep doubling 42 times. You know 1>2>4>8>16 etc. (the miracle of compound interest).

So we need to keep the genes that are to be expressed later in the development program shut down until it is the right time for them to be turned on, because if we turned on all the genes in the DNA on at the same time it would be complete chaos and certain death.

So how are genes kept shut off until it is the right time for them to be expressed? That is where the methylation comes into play. So now let us modify our string of genes we made above and modify them so that only the first gene is being expressed.

We would shut off all the other genes other than Gene1 with methylation. What is methylation? It is just the attachment of a methyl group to the beginning of the gene to cover it up so it cannot be read by the gene reading machinery. (Think of how a cassette tape is played –the DNA/genes are the tape.)

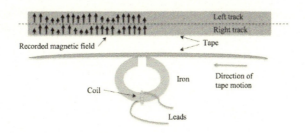

Let us use M as a symbol for a methyl group. (By the way a methyl group is just a tiny little molecule with a carbon in the middle with 3 hydrogens attached to it looks like this:

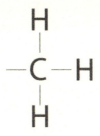

The little extra line sticking out is the bond where it can attach to the beginning of the gene. Okay, so we will just represent a methyl group with the letter M, now let us shut down all the genes on our DNA except GENE1

      M   M   M

Gene1--Gene2--Gene3--Gene4-

So this piece of DNA will only give out the instructions encoded in Gene1 which we earlier said would cause the single cell egg to divide into two cells.

Now after the cell has divided into two cells we now want to activate the second gene, Gene2 that tells the 2 cells to become 4 cells...so we just remove the second methyl group

        M   M

Gene1--Gene2--Gene3--Gene4-

So you see here how development is controlled by the sequential expression of genes which remain shut off by methyl groups covering them up until it is the right time for them to be turned on.

So that's it; that's as far as I am going to go to have you understand a very simple model of DNA methylation controlling the timing of gene expression.

So what does this tell you from a BIG PICTURE point of view? The age of an organism is dictated by how many methyl groups are left on the DNA. Once most of your methyl groups are gone, most of the genes have been expressed and you have become a fully grown adult. In fact the level of DNA methylation decreases with each round of cell division, and human cells are generally limited to 50 to 70 divisions in a laboratory dish before they die. It has been shown that the level of DNA methylation in a cell corresponds quite accurately to how many times it can double before death. The less methylation, the fewer doublings possible, or another way to put it, the less DNA methylation a cell has, the older it is. Yes the level of DNA methylation is like a clock that can tell you how much time you have left to live!

Methylation and transcription MBD

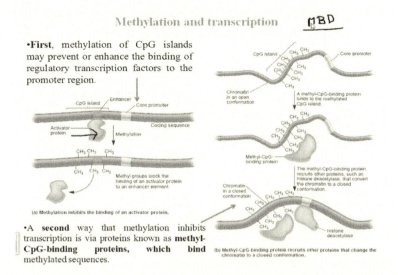

- **First**, methylation of CpG islands may prevent or enhance the binding of regulatory transcription factors to the promoter region.

- A **second** way that methylation inhibits transcription is via proteins known as **methyl-CpG-binding proteins**, which bind methylated sequences.

(a) Methylation inhibits the binding of an activator protein.

(b) Methyl-CpG-binding protein recruits other proteins that change the chromatin to a closed conformation.

Now this is highly oversimplified but as a general overall principle it is valid; I'll give you some evidence to back this up in a minute. But let's now ask what happens after we become an adult? Is that the end of any additional genes being expressed by the development program?

If you ask the selfish gene theorists they would likely tell you yes-the development program keeps changing which genes are expressed until you have become and adult, and then after that it stops! They would say there are no more additional developmental "good" genes expressed after you become an adult. And certainly there would be no "bad" aging genes expressed because "they don't exist"! They believe that aging is caused by damage to the DNA, not by the expression of aging genes.

Their view has to be 100% wrong if there exists aging genes which when activated give out instructions to your body to self-destruct! From the perspective of the BIG PICTURE, another day older is another day older as far as your genes are concerned regardless if you are a baby, a 40 year old, or a 75 year old! When you are younger, becoming older is a good thing, but there comes a point where another day older means another day closer death. Another day older means aging genes are being activated whose sole purpose is to kill you or at least to stop you from reproducing. So let's take a look further along the DNA strand we constructed earlier:

<pre>
              M   M

Gene100 ----AGene101---- AGene102----AGene103
</pre>

What we show above is the point at which the last gene that is good for you has been expressed (GENE100) and the first aging gene has been activated by loss of methylation (Agene101), and aging genes 102 through 105 are all lined up waiting to get activated one by one as time goes on as soon as they become activated when they lose their methyl groups (M's).

So is this really a correct way to look at aging? Is all you have to do stop the aging process is to keep your methyl groups from coming off? "Keep your bumps on" as I once explained to a friend of mine. I think the answer is mostly a yes but it gets a bit more complicated but not that hard to understand. I'll pursue this more shortly. How about the evidence?

In 1997 when I was writing my firs paper on aging, I was trying to understand how DNA methylation was involved in the development and aging process. There was quite a bit of information regarding how dramatic changes in DNA methylation were involved in the maturation of a fertilized egg into an embryo and then a fetus and eventually a human baby.

The process was described generally as follows:

Upon fertilization of the egg by the sperm, the entire length of the DNA is completely demethylated (all the methyl groups are removed).

The next step is the entire strand of DNA is heavily re-methylated.

Then each time there was a cell division e.g. from 1 to 2 cells, 2 to 4 cells, 4 to 8 cells, etc. There was a dramatic loss of methylation causing many various genes to become activated. As a cell gets further and further along towards achieving their final identity, each type of cell acquires a specific methylation pattern that activates just the genes needed for that cell type while keeping other genes useful in other kinds of cells repressed. This is known as differentiation-which basically means causing different genes to be expressed in different cells.

If you re-methylated (shut down) all the genes that are expressed in say a skin cell, you could turn that cell hypothetically back into a single cell fertilized egg. This is exactly what happens when you clone an animal from a cell from its body. The DNA is removed from a fertilized egg and it is replaced with the DNA you take out of an animal's (say) skin cell. You give the egg a shock and it re-methylates the DNA of the skin cell and voila you have a brand new baby clone exactly identical to the donor animal.

Getting back to the development program we were just discussing, after a rapid loss of methylation as the embryo grows larger, eventually the de-methylation of the DNA slows down to a crawl, and remains at a much slower pace for the rest of the developing human or animal's life.

So other than cell division, is there any other way to influence how quickly we lose our DNA methylation? Another prediction I made in that 1998 paper was that yes there should be. The changes in various hormone levels over our lifetimes influence how quickly we lose methylation or even regain methylation (become rejuvenated!). And yet again evidence is coming to light that hormones do indeed affect methylation by altering the activity of DNA methyltransferases.

I didn't mention DNA methyltransferases (DNMT's) yet , but quite simply they are the little machines floating around in your nucleus that take methyl groups and attach them to your DNA at various places. A 2009 study showed that

"Recent results suggest that DNMT expression is under hormonal control. For instance, DNMT3A, DNMT3B, and DNMT1 are under the regulation of female sex steroid hormones during the menstrual cycle (Yamagata *et al.* 2009, van Kaam *et al.* 2011)".

Research is popping up all over and even showing some of the "good" hormones like melatonin actually drive the re-methylation of your DNA, for example:

British Journal of Cancer (2013) 108, 2005–2012. doi:10.1038/bjc.2013.188 www.bjcancer.com
Published online 30 April 2013

"Melatonin-induced methylation of the ABCG2/BCRP promoter as a novel mechanism to overcome multidrug resistance in brain tumour stem cells."

So the simple (overly simple) way to look at DNA methylation at this point is that in general loss of DNA methylation drives development and eventually aging via sequential gene expression. And that the good hormones that decline with age slow or reverse the aging process by promoting re-methylation of your DNA, while the "bad" hormones that increase dramatically after age 40 in humans accelerate the aging process by de-methylating your DNA.

So back in 1997-1998 this was a pretty radical idea, but the proof is coming out that it is for the most part correct as an overarching principle of aging. Back then I thought DNA methylation was the be all and end all factor that controlled aging. But I was only part-right. It turned out there were also other things that also controlled aging by suppressing the expression of your genes by covering them up, but they weren't covered up by DNA methylation. However, they all share the same concept of suppressing aging and development genes by covering them up until the time is right for their expression. What are these other things?

-Chromatin Condensation/Histones

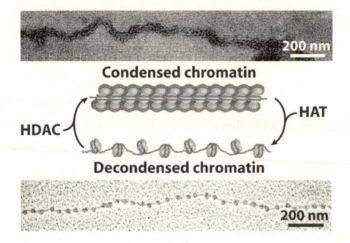

Chromatin condensation is a very similar concept to DNA methylation. Chromatin is just another name for DNA that has been tightly wrapped up and covered with proteins so that any genes in the DNA cannot be turned on. The DNA is tightly coiled around little spools called histones and also covered by various other proteins which act like the insulation covering a wire. Aging genes are suppressed by keeping these genes suppressed and covered up. When the histones are acetylated, they open up allowing access to these genes. Thus compounds that interfere with this opening up of the DNA are good for you and can slow the aging process they are known as histone deacetylases. A well-known class of these histone deacetylases are known as SIRTUINS which stands for Silent Information Repressors. They should actually have been named AGRTUINS or aging gene repressors. The health supplement resveratrol acts by stimulating the expression of the SIR1 gene which then silences aging genes on the DNA by promoting chromatin condensation/gene silencing.

Telomeres:

Telomeres are actually stretches of DNA that are found at the ends of each chromosome. (A chromosome is just 1 of 23 segments of DNA that exist as strands, kind of like one book in an encyclopedia set where the encyclopedia set represents all your DNA). Telomeres also function in the same manner as DNA methylation to suppress aging genes. This is not well known yet, so let me explain.

Telomeres do not have any genes on them, you can think of them as just junk DNA. It is thought by many that once the telomere shrinks enough then the genes next to it will be damaged and cause aging. This is wrong. Actually telomeres are highly controlled gene regulating devices. After a cell and its chromosomes divide and the telomere shortens, the telomere then folds back over onto the DNA that has genes on it, and it suppresses these genes by covering them. This is known as the telomere position effect. As the telomere shortens with each cell division, in a very controlled and measured way, it exposes aging genes one by one. So in effect telomeres are doing the same thing as DNA methylation and chromatin condensation –suppressing aging genes. And it should come as no surprise that telomeres can be re-lengthened by an enzyme called telomerase, and that telomerase is under hormonal control.

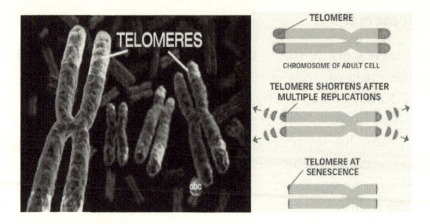

Rethinking Telomeres

Not only do telomeres protect the ends of chromosomes, they also modulate gene expression over cells' lifetimes.

By Kate Yandell | March 1, 2015

REGULATORY ROLE: Early in life, when telomeres (red) are long, chromosome looping brings them into contact with particular genes (green) (1). As cells age, their telomeres shorten. Through mechanisms that are not yet understood, this alters chromosome looping and telomeres' interactions with genes, leading to age-related changes in gene expression (2). Imaging using 3D-FISH (right panels) illustrates the distance between a certain gene and long (top) and short (bottom) telomeres.

(NOTE-the mRNA being produced by the DNA in picture 2 indicates gene expression is occurring in the presence of the shortened telomere).

Lamin A proteins:

The most interesting fact I learned after the publication of my first aging paper was that the rapid childhood aging disease of progeria is also caused by a defect in a protein that is involved in silencing genes, the lamin A protein (interestingly it is Animal spelled backwards! And combined it spells Animal lamina-which is interesting because lamin A protein actually does act like a laminate, covering the nuclear membrane).

The selfish gene theorists rushed out to say that progeric lamin A was a protein defect which resulted in DNA damage and that caused "accidental" aging just like the accidental DNA damage that they say causes aging in the normal adult. But this again, as usual turns out to be wrong. In addition to providing the covering for the nuclear envelope that houses the DNA, the lamin A proteins also affect placement of various proteins and molecules that CONTROL/SUPPRESS gene expression. So what is really going on here is that the aberrant lamin A proteins are failing to suppress various aging genes that cause an aging program to kick in at an early age which leads to the rapid aging disease of progeria.

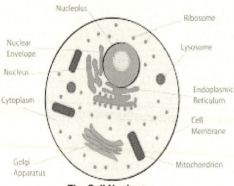

The Cell Nucleus

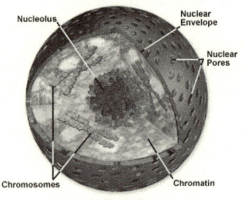

Figure 1

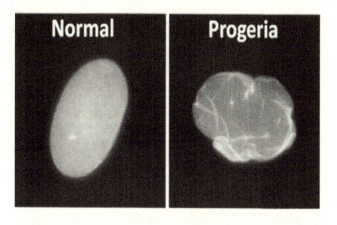

Lamin A: Normal Protein vs. Progeria Protein (aka progerin)

Press release by Evan Lerner, University of Pennsylvania

Now, a study by researchers at the University of Pennsylvania has shown that a protein found in the nuclei of all cells — lamin-A — plays a key role in the differentiation process.

"We show that lamin-A is regulating gene expression through epigenetic mechanisms," Ivanovska said. "It can catch certain transcription factors inside the nucleus."

Transcription factors are proteins that move in and out of the nucleus to trigger protein production. Some of the transcription factors that lamin-A controls relate to a key vitamin A metabolite called retinoic acid, but this also regulates lamin-A production. This suggested a feedback loop that helps to drive increased nuclear stiffness.

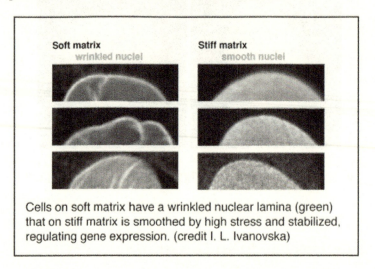

Cells on soft matrix have a wrinkled nuclear lamina (green) that on stiff matrix is smoothed by high stress and stabilized, regulating gene expression. (credit I. L. Ivanovska)

So the bottom line about lamin A proteins is that they work the same way DNA methylation, chromatin condensation and telomeres all work, by suppressing aging genes. It seems once evolution has found a working concept, it uses it over and over again.

Do lamin A proteins play a role in normal aging? Yes, it seems they do, as it turns out that the same defect that childhood progeria patients get in their lamin A proteins also starts showing up in normal adults at older ages.

From: Eur J Hum Genet. 2009 Jul; 17(7): 928–937.2009 Jan 28. doi: 10.1038/ejhg.2008.270

Increased expression of the Hutchinson–Gilford progeria syndrome truncated lamin A transcript during cell aging

"It is not clear how HGPS relates to normal physiological aging, but the identification of mutations in the *LMNA* gene has increased interest in the search for the role of the lamin proteins in the aging process. Studying HGPS cells, Goldman *et al*[14] showed that nuclear defects accumulate as the cells become older in culture, which is accompanied by an increase in the amount of progerin. Progressive alterations in nuclear architecture also accompany aging in the nematode, *Caenorhabditis elegans*.[19] Scaffidi and Misteli[20] showed that cells from normal individuals of old age have similar defects compared with to those reported in cells from HGPS patients. This study also showed that the cryptic splice site, which is active in HGPS cells, is also used in cells and tissues from both young and old healthy individuals, and inhibiting the splice site in cells from old individuals reversed their nuclear abnormalities.[20] Additional evidence implicating progerin in the normal aging process comes from recent reports that show that the number of progerin protein positive cells increase with age, both *in vitro* and *in vivo*, in samples from apparently healthy individuals."

I have a good hunch, which I might share with you later, that elevated FSH levels are involved with triggering the expression of the aging genes that are also expressed prematurely in progeria. It is becoming QUITE CLEAR that progeria is just the acceleration of one segment of the program of normal human aging (what I call the "male" segment).

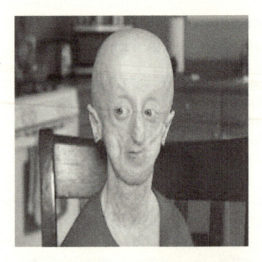

WRN Proteins: I must admit, since around the year 2000 when the last of my three scientific papers on aging and evolution were published, I haven't really been keeping up that well with the recent research into aging. It really didn't help motivate me when the predictions I made in the 1998 paper kept coming true. I thought I had it all figured out. So I was lazy. While doing some research for this book, I decided to do some searching for any new information about Werner's disease. The rapid aging disease that causes normal teenagers to age into the equivalent of a 90 year old person by age 45 or so.

Guess what?-It turns out that the protein that is defective in Werner's disease called WRN, also plays a role in silencing various aging genes! This was a bit of a crazy idea to most, because the WRN protein is what is known as a DNA helicase, which means that it has a duty of attaching to the DNA and then causing it to wind or unwind during gene activation and DNA copying. However, like lamin A it turns out that WRN also has a second duty. It suppresses various genes in stem cells, and it also suppresses aging genes that cause inflammation and generate fat cells during aging and various other aging related genes, 14 in total! It likely does this by interacting with and keeping the chromatin condensation silencing of genes maintained and organized, as cells from Werner's patients, like the cells in progeria patients, have disorganized chromatin which is involved in silencing genes (like insulation covering a wire). So once again, evolution has used the same playbook of silencing aging genes one more time.

DNA disorganization linked to aging

Study of Werner's gene protein (WRN) shows that resulting cell chaos leads to graying hair, brittle bones

BY
RACHEL EHRENBERG
2:00PM, APRIL 30, 2015

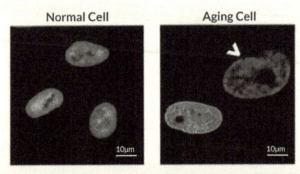

NUCLEAR DISARRAY Normally tightly packed bundles of DNA (as in the normal stem cells, left) come undone in cells that age prematurely (right), enlarging their nuclei and promoting aging.

In addition to silencing various aging genes, WRN is also involved in some manner with DNA methylation of DNA and telomere length maintenance.

Another interesting study also came out in 2010 that showed WRN is controlled by estrogen! Giving estrogen to a line of female breast cells causes them to produce WRN protein, while testosterone does not have this effect. This is especially interesting because Werner's Disease doesn't kick in until puberty which likely means that the WRN protein evolved to control genes that are expressed after childhood, pubertal development genes, as well as aging genes. This also likely means that age-related hormone changes also control the long term decline of WRN levels with age which were documented in a recent study.

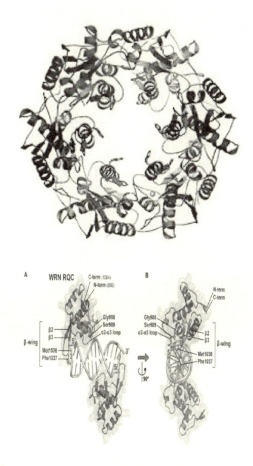

A Diagram Of How 6 WRN Proteins Come Together

To Create A DNA Helicase Complex That Winds And Unwinds DNA-

The Space In The Middle Of The Complex Is Where The DNA Localizes.

A Single WRN Protein –Apparently single WRN proteins have anti-aging duties

CHAPTER 11: How Sex And Aging Are Selected For- Everywhere & All The Time

This chapter will explain the biggest problem that selfish gene theorists have with the idea of programmed aging and aging genes.

Even if aging could evolve in the short term, even accidentally, they wonder how could it be possible that evolution can select FOR programmed aging and aging genes in the long run?

It seems so obvious to selfish gene theorists that evolution, driven by the selfish gene, should quickly find ways to rid the host of the aging genes and the aging program to enable increased reproduction and wider spread of the selfish genes. To them it is just not possible for evolution to select FOR something that is bad for you and slows the spread of your genes!

There have been attempts in the past to counter this objection by suggesting that evolution selects for aging, even though it is bad for the individual, because it is <u>good for the group</u> in some way.

This idea is what is known to biologists as <u>group selection</u>. And as mentioned before, back in 1966 George C. Williams wrote a book making a laughing stock of the idea. His book led to a long period where anyone in the biology community was scared to even mention the term. The reason I say he set evolutionary biology back 50 years is because of how badly he disparaged the idea of group selection that he discouraged most young biologists from even thinking about any form of multi-level selection for almost a generation. And eventually Williams saw the error of his ways and in 1992 admitted in his book *Natural Selection: Domains, Levels, and Challenges*, that he had finally accepted the idea of a higher level of evolutionary selection in clade selection which is basically selection between competing branches of the evolutionary tree, each branch containing multiple species-a much higher level of selection than group selection. But this was too little, too late, and did not undo the damage his earlier musings had caused.

To this day the great majority of evolutionary biologists believe that group selection occurs only in the most rare of cases such as the example of the beehive full of non-breeding individuals and a single reproducing queen. They believe this species to be an oddity that can basically be ignored.

Now don't let me make you think that I believe in group selection as defined by current mainstream theorists. I also think it is usually an unimportant force of evolution based on their definition which I will describe shortly.

A few theorists including E. O. Wilson recently (2010) have argued that multi-level selection (their rehabilitated term for group selection) is important to make up for deficits they saw in the explanatory power of the selfish gene. But for every evolutionary biologist suggesting that multi-level selection is an important force of evolution , there may be about one hundred who say it is not.

So what were the details about this "new" idea of multi-level selection? It simply suggested evolution can operate at different levels, the lowest level is the **genes, next come the cells, then the organism** level and finally the groups. It was suggested by a few that evolution can operate at all these levels.

From the BIG PICTURE point of view developed herein- None of these levels of selection can explain the existence of sex and aging! They are missing the most important level. One needs to go up even one more level to get to the proper selection force that selects for sex and aging and that is SPECIES-SELECTION.

Now let me admit one thing right here. In my first paper on aging from 1998, I tried to explain the existence of aging as a group-selected trait. I did this by creating an example of two kinds of rabbits that lived on two sides of an island separated by a high, impassable mountaintop. An evolving predator was able to visit both sides and eat the rabbits. I constructed the scenario so that the aging rabbits had greater genetic and phenotypic diversity. (Phenotypic just means what kind of actual body you have that is created from your genes. It is the physical manifestation of the output of all your genes in the real world). Under this scenario the aging rabbits had a few fast members that were able to survive and reproduce, while the identical non-aging (and asexual) rabbits were all the same and all got eaten and went extinct.

One day I realized that this was a ridiculous scenario and that group selection between two potentially interbreeding groups would eventually come undone if the two differing groups ever met and interbred. Interbreeding would quickly destroy their differences.

It was then a light bulb went off in my head! The group selection I was thinking of would be two groups that compete in the same local ecosystem for the same resources, and they could not interbreed! This ruled out two groups of rabbits, but it ruled in two groups of competing, non-interbreeding species like rabbits and ground squirrels! Or in other words it wasn't group selection that was selecting for sex and aging, but rather species selection at the local level! I corrected the error I made in my 1998 paper in a paper published in 2000 titled Sex, Kings, & Serial Killers and Other Group Selected Human Traits. At the time I published it I did not know there was a term for species selection, and I assumed that the biology community had just created an incorrect definition of group selection. In that paper, I corrected the term to be defined as selection between two groups of non-interbreeding species that competed at the local level of the ecosystem. Little did I know that once in 1975, and then again in 1993 biologists had started using the term species selection, but it was hardly on anyone's radar. In fact as of the date of this writing, 2015, the term species selection can only be found 11 times in the science data base Pub Med. And the first time used by Stephen Jay Gould was 1993. So excuse me if I did not know the term existed and just assumed that biologists had an incorrect definition of the form of group selection that actually worked in nature.

Now, knowing the term exists, the correct title of my 2000 paper should have been: Sex, Kings, & Serial Killers and Other ~~Group~~ **SPECIES** Selected Human Traits

Here is the complete list of the 11 science papers that mention the term species selection in the title as of today's date (2015):

(none of them suggest that species selection is the force that selects for sex and aging).

A theory of evolution above the species level.
Stanley SM.
Proc Natl Acad Sci U S A. 1975 Feb;72(2):646-50.

Species selection on variability.

Lloyd EA, Gould SJ.
Proc Natl Acad Sci U S A. 1993 Jan 15;90(2):595-9.

Species selection on organismal integration.
Björklund M.
J Theor Biol. 1994 Dec 21;171(4):427-30.

A genetical theory of **species selection**.
Rice SH.
J Theor Biol. 1995 Dec 7;177(3):237-45.

Developmental shifts and **species selection** in gastropods.
Duda TF Jr, Palumbi SR.
Proc Natl Acad Sci U S A. 1999 Aug 31;96(18):10272-7.

Between- and within-host **species selection** on cytoplasmic incompatibility-inducing Wolbachia in haplodiploids.
Vavre F, Fouillet P, Fleury F.
Evolution. 2003 Feb;57(2):421-7.

Punctuated equilibrium and **species selection**: what does it mean for one theory to suggest another?
Turner D.
Theory Biosci. 2010 Sep;129(2-3):113-23. doi: 10.1007/s12064-010-0088-6. Epub 2010 May 26.

Correlations of life-history and distributional-range variation with salamander diversification rates: evidence for **species selection**.
Eastman JM, Storfer A.
Syst Biol. 2011 Jul;60(4):503-18. doi: 10.1093/sysbio/syr020. Epub 2011 Apr 2.

Species selection and the macroevolution of coral coloniality and photosymbiosis.
Simpson C 2013 Jun;67(6):1607-21. doi: 10.1111/evo.12083. Epub 2013 Apr 9.

Evolution. 2013 Jun;67(6):1607-21. doi: 10.1111/evo.12083. Epub 2013 Apr 9.

Species Selection Favors Dispersive Life Histories in Sea Slugs, but Higher Per-Offspring Investment Drives Shifts to Short-Lived Larvae.
Krug PJ, Vendetti JE, Ellingson RA, Trowbridge CD, Hirano YM, Trathen DY, Rodriguez AK, Swennen C, Wilson NG, Valdés ÁA.
Syst Biol. 2015 Nov;64(6):983-99. doi: 10.1093/sysbio/syv046. Epub 2015 Jul 10.

The extended Price equation quantifies **species selection** on mammalian body size across the Palaeocene/Eocene Thermal Maximum.

Rankin BD, Fox JW, Barrón-Ortiz CR, Chew AE, Holroyd PA, Ludtke JA, Yang X, Theodor JM.
Proc Biol Sci. 2015 Aug 7;282(1812):20151097. doi: 10.1098/rspb.2015.1097.

So-Boom! Bada Bing! Species selection explains how aging and sex are selected for EVERYWHERE all the time all over the world and it is a very subtle, but strong force of evolution. It is as easy as considering two competing groups, let's stick with rabbits and ground squirrels. They both compete for the same territory and resources. Assume one of them reproduces sexually, and ages, of course we will assign these traits to the rabbits, while the ground squirrels do not age and they reproduce asexually. What happens when an evolving predator shows up? If the predator can kill one of the asexual clones, he can quickly kill them all and the non-aging asexual ground squirrels will quickly go extinct at the local level.

Because the rabbits are a much more diverse lot, there will be fast ones, slow ones, smart ones, dumb ones, careful ones, bold ones, black ones, grey ones, etc. When confronted with the evolving predator, it is much more likely that a mating pair of the aging/sexual rabbits will survive compared to the ground squirrels. And you know how fast one pair of rabbits can multiply!

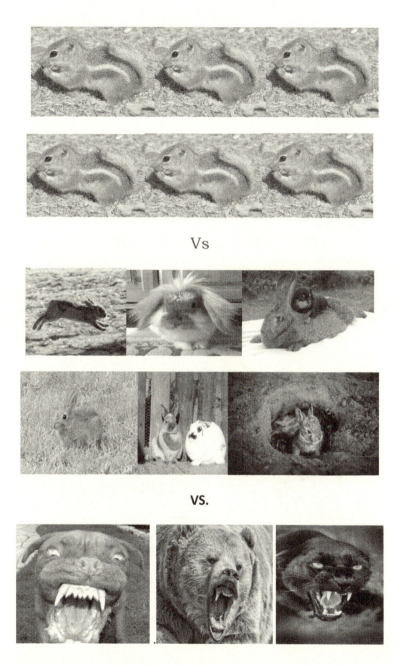

Vs

vs.

Most evolutionary biologists believe that selection at the species level is impossible, but there was one, the late

Stephen Jay Gould who argued for its possible existence. Although Gould championed the idea of species selection he had no real examples of its existence nor did he describe in detail how it might work, rather he just noted that some species might have evolved some traits that prevent them from going extinct, while others do not, and even if these traits were not helpful for enhancing an individual's spread of his or her selfish genes, the traits might evolve and be conserved by evolution as a defense to extinction. Gould seemed mostly preoccupied with mass extinctions caused by global catastrophes and within this framework he toyed with the idea of species selection.

Two quotes about Gould's views on species selection are:

When referring to mass extinction events:

"species with broad geographical ranges, species with broad habitat tolerances, species whose lifecycle does not tie them too closely to a particular type of community all would have had a better chance of making it, (p. 110) and this amounts to species selection. However, there are no well-worked-out case studies."

"So it has been hard to find really convincing examples of species-level properties that are built by species-level selection. The problem is to find: (i) **traits that are aspects of species, not the organisms making up the species**; (ii) traits that are relevant to extinction and survival; and (iii) traits that are transmitted to daughter species, granddaughter species and so forth". And "transmission to daughter species is especially problematic".

I believe Gould is almost on the right track here, that yes, species selection refers to traits that are relevant to extinction and survival and to traits that are transmitted to daughter and granddaughter species.

However, I think a revision is necessary for "traits that are aspects of species, not traits of the organisms that make up the species". Instead I believe the working definition will be found by looking for traits that the species in competition with one another have in common and are not traits unique to the individuals of the species.

In this case, species selected traits would include sexual reproduction, male and female sex types where males tend to attract predator attention, and aging. And they would not include things unique to a particular species such as say, leaves and roots, number of toes, types of food preferred, body plan, number of arms and legs, etc.

The species selected traits of sex, male and female sex types, and aging are all relevant to extinction (caused by predation) and survival at the local level, and all can be passed onto daughter species.

So if I have uncovered the correct way to look at species selection which allows it to be a major force of evolution, where did Gould's view of species go off track? It seems Gould fell into the trap of looking at most extinctions as being one big decline of the species all over the world in a short period of time. He was focused on major catastrophes that wiped out large numbers of species all over the earth in a short period of time.

My view of extinction is quite different and I believe is the correct way to look at extinction to see how it fits in with the way evolution usually works.

Extinction is usually not a single big event at the world level, but rather it is an event that occurs all the time at many different local levels. Try to imagine how difficult it is to have a rapid global extinction of a species. The extinction would have to occur in every local ecosystem all over the world where the species lives wiping out every breeding pair on earth so they could not rebound. If just a single breeding pair was left in a single local ecosystem, (or maybe just a single female who can have a virgin birth of sex changing females) extinction will not occur, as the couple or female can repopulate their local ecosystem with offspring, and these can later migrate out to other ecosystems to reestablish the species on a large scale.

From the BIG PICTURE perspective, extinction is a local event that occurs in a local ecosystem. If all the ground squirrels are killed in say Yellowstone National Park, and it is not immediately connected to other small ecosystems, say maybe because it is surrounded by desert, then you can say that ground squirrels are extinct in Yellowstone Park. It is not relevant for our purposes that ground squirrels exist elsewhere on the planet. Once ground squirrels go extinct in Yellowstone Park, the niche they occupied in Yellowstone can then be occupied by any of its competitor species, let's say rabbits.

So remembering that we earlier made our ground squirrels asexual, we can imagine that they went extinct in Yellowstone Park due to the introduction of a predator , say wolves, that were able to find and kill them all since they were all identical.

The wolves were not able, however, to kill all the sexually reproducing rabbits due to their much higher level of diversity. A predator-defended mating pair was able to survive and then expand the population to rebound and also to occupy the niche left empty by the loss of the ground squirrels. Eventually, these aging and sexually reproducing rabbits will have a chance to migrate to other local ecosystems and compete for territory left open due to the local extinction of any non-aging, asexual rabbit populations that went extinct.

The bottom line is that evolving predators all over the world are causing asexual, non-aging prey species to go extinct all the time which allows only aging, sexual species to survive locally and then migrate to other ecosystems. Over time this process leads to the situation where almost all animals in all ecosystems are sexual reproducers who age. Or in other words, **sex and aging are selected for almost all the time everywhere on the planet where there exist evolving predators.**

It is obvious as to how sexual reproduction would lead to increased diversity in a prey species due to the rapid mixing and rapid recombination of genes with every generation. One only needs to think of the case where two separate unrelated beneficial mutations occur in two different individuals. In the case of asexual reproducers, these two beneficial mutations might never ever come together in one individual unless one of the carriers of a beneficial mutation undergoes a second beneficial mutation. What are the odds of that-slim? It would take a long time. Now in the case of sexual reproducers, if the two carriers of the two separate beneficial mutations were to mate or any of their offspring to mate, (and who were more likely to survive) it would not be too long before the two beneficial mutations would come together in the same individual.

But how does aging increase the diversity in a population?

There are two ways to look at it. And both involve limiting the contribution of any one individual to the gene pool. This can be done in one of two ways, or both ways. The first way is for aging to reduce the amount of offspring one has and this is no more evident than when you consider human female menopause. Basically menopause puts a cap on a woman's contribution to the gene pool. A woman can start having children at say age

15, and continue until about age 42 (where most women have trouble conceiving prior to the full blown menopause that kicks in around age 50). Keep in mind that breast feeding acts as a form of birth control for about 6 months where the woman can no longer get pregnant. So we can adjust the equation as follows: 42-15 = 27 years of reproduction potential and 15 months are required to get pregnant and breast feed per baby. So 27 x 12/15 = 21.6 children maximum per woman per lifetime.

If not for menopause, if a woman could continue reproducing until say age 80, the number of children she could have had would be 80-15 x 12/15= 52!

World Record-Oldest Mother to Give Birth (from donor eggs)

Omkari Panwar – 70 Years Old

In a desperate mission to have a son and give her husband an heir, Omkari Panwar became the oldest mother in the world when she became a mother to twins. In 2008, in the hospital of Muzaffarnagar (about a seven hour drive from New Delhi), she gave birth to twins prematurely, a boy and a girl, both weighing about two pounds. Pamwar and her husband

Charan already had two adult daughters who had children of their own, but the couple were desperate to have a son. Panwar went through in-vitro treatments, and stated that pregnancy at such a young age was not only more difficult, but much more painful as well. To pay for the IVF, Panwar's husband sold the family buffalo, took out a credit card loan, spent life savings, and mortgaged his land.

So evolution could either use menopause to stop a woman from reproducing, or it could simply kill her at a certain age which is what happens in many other animals. Most nonhuman animals have females that experience declining litter sizes as she ages, and then when the litter size drops from 1 to 0 is usually the same time that she dies as well.

The British woman who became the world's oldest natural mother at 59

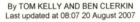

By TOM KELLY AND BEN CLERKIN
Last updated at 08:07 20 August 2007

A British housewife became the world's oldest natural mother after giving birth at the age of 59, it was claimed yesterday. Dawn Brooke had a healthy boy without any fertility treatment only 12 months before she became eligible for her old age pension, her family said.

Now when we consider males of all animals, we see that their reproduction is almost unlimited which makes sense if males are merely vehicles for predator testing of genes. The few surviving males should be allowed to spread predator tested genes far and wide, but evolution also puts a limit on them too by killing them at a certain age. In most animals this is about the same age that the female becomes infertile/dies of old age.

What might be the upper theoretical limit for how many children a human male could sire in a lifetime? Again let's start at age 15 to begin with and end at age 90. And let's say our father trains hard and could develop the habit of being with three different women per day (it's good to be the king!). Let's say one eighth of the encounters resulted in a pregnancy, we would end up with:

(90-15) x 365 / 8 = 3,421 children! Quite an increase compared to a human females paltry 21.6! One could say from this exercise that men are 158 times more fertile than woman, which says a lot for the men who practice lifelong monogamy!

The oldest ever man to father a child was reportedly Les Colley (1898 - 1998, Australia), who had his ninth child a son named Oswald to his third wife at the age of 92 years 10 months. Colley met Oswald's Fijian mother in 1991 through a dating agency at the age of 90. "I never thought she would get pregnant so easy, but she bloody well did," he told newspaper reporters.

World's oldest father has 21st child at 90

August 2007

The world's oldest father has done it again, fathering a child for at least the 21st time, at the age of 90.

Indian farmer Nanu Ram Jogi, who is married to his fourth wife, boasts he does not want to stop, and plans to continue producing children until he is 100.
Mr. Jogi admits he is not certain how many children his series of four wives have borne him - but counts at least 12 sons and nine daughters and 20 grandchildren.

Two-week-old daughter Girija Rajkumari is the latest addition to the proud father's family. "Women love me," Mr. Jogi said. "I want to have more children. I can survive another few decades and want to have children till I am 100 - then maybe I will stop."

Now another more complicated way to look at this, is that evolution, if it favors diversity in the various combinations of genes, and body types (phenotypes), it will favor a younger fertile population as opposed to an older fertile population. Why is this so?

As you look at older and older age groups of a population, there will be fewer and fewer individuals. This is because over time the longer one lives the more likely one will be killed by a disease, an accident, a predator, etc. So if you look at the oldest fertile group of individuals in a population you will see that the few surviving members of this group have all survived the same series of selection events and thus are much more similar to each other than younger members of the population are who have not been winnowed down by a long series of deadly incidents.

Thus possibly all the oldest members that have the ability to survive an epidemic, have also all survived a fire, survived being hunted by say lions, survived starvation, etc. So after each mortality event, the population becomes less diverse.

Below is the graph created by Peter Medawar in his famous paper on aging (An Unsolved Problem of Biology) to show the age distribution of a population of test tubes that experience a constant rate of breakage over time. You see that there are many more younger test tubes than older ones.

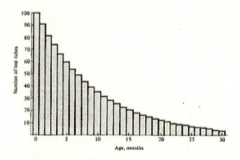

High Phenotypic Diversity >>>   Lower Phenotypic Diversity

The insight that I will add to his age distribution of test tubes is that if you are dealing with biological populations, the younger individuals will be more diverse in gene combinations and body variations than the older individuals because they have not yet experienced a long series of homogenizing selection events.

The concept is similar to separation of a starting mixture by pouring it through a column full of filtration gel, everything gets separated based on weight, and while the beginning mixture is quite diverse, what is left at the end of the journey through the gel is a group of molecules of all the same kind. So just imagine the direction down the column is time, and whenever a molecule gets caught in the gel it is the same as a group of vulnerable individuals being killed by a selection event.

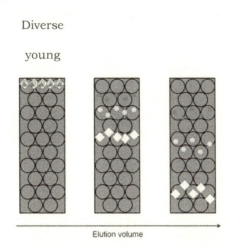

Diverse young

Elution volume

Elderly survivors

The bottom line is if evolution favors diversity it will evolve mechanisms to maintain a population skewed towards younger individuals (aging). If the environment doesn't do this through various deadly trials (predation), evolution can also enforce this population distribution by killing off the members that live too long via programmed aging. I used to call this aspect of aging- the predator's baby sitter as aging takes over in maintaining the youth skewed population distribution in times when the predator is not around to do it.

The science world needs a young math modeler to try and recreate the following example on a computer:

It should seem pretty obvious to all who have read this chapter up to this point that an evolving predator can select, at the local level for sexually reproducing and aging organisms by killing all of those organisms who lack enough diversity to quickly evolve a defense to their evolving offense.

This of course oversimplifies the situation. The situation is infinitely more complex, as almost every evolving predator is also prey to another species. And every prey species, other than plants, is also an evolving predator, if not of other animals then of plants or seeds.

So what we get in every local ecosystem is a huge complicated simultaneous equation with a huge number of interrelationships.

And what I have surmised from this big interacting ball of evolving equations, is that sex and aging are basically emergent traits that evolve out of the whole big mess. One way I tried to explain it to someone once was that sex and aging are selected for because every organism in an ecosystem is in competition with all the other organisms either directly or indirectly to claim a portion of the biomass for their existence. So the elephant has evolved aging and sex because the lion has evolved aging and sex because they compete with ants who evolved aging and sex, who compete with plants who evolved aging and sex, etc., etc. And no matter how different two species can be from each other, in most cases the one thing they all have in common is aging and sex- truly a species level trait. If one doesn't know what to look for, it would be easy to miss.

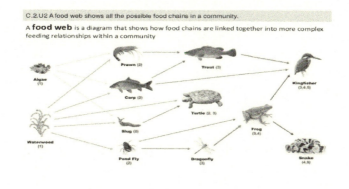

C.2.U2 A food web shows all the possible food chains in a community.
A **food web** is a diagram that shows how food chains are linked together into more complex feeding relationships within a community

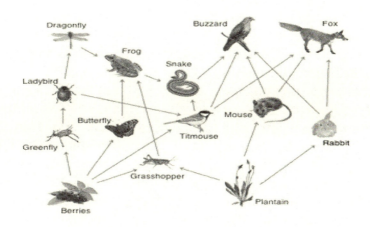

FROM ALL THIS EMERGES SEX AND AGING

I have begged all the math modelers I know to try and simulate this idea in a computer model, but I still, as of yet have no takers. The closest thing that has been created to date is a math model by Andres Martins who created the following math model on his own, but who so far has ignored me after I told him about my idea and am still waiting for him to give it a try. Here was a math model of his that is definitely on the right track but still lacking:

Change and Aging Senescence as an Adaptation

- André C. R. Martins

Published: September 16, 2011

Abstract

Understanding why we age is a long-lived open problem in evolutionary biology. Aging is prejudicial to the individual, and evolutionary forces should prevent it, but many species show signs of senescence as individuals age. Here, I will propose a model for aging based on assumptions that are compatible with evolutionary theory: i) competition is between individuals; ii) there is some degree of locality, so quite often competition will be between parents and their progeny; iii) optimal conditions are not stationary, and mutation helps each species to keep competitive. When conditions change, a senescent species can drive immortal competitors to extinction. This counter-intuitive result arises from the pruning caused by the death of elder individuals. When there is change and mutation, each generation is slightly better adapted to the new conditions, but some older individuals survive by chance. Senescence can eliminate those from the genetic pool. Even though individual selection forces can sometimes win over group selection ones, it is not exactly the individual that is selected but its lineage. While senescence damages the individuals and has an evolutionary cost, it has a benefit of its own. It allows each lineage to adapt faster to changing conditions. We age because the world changes.

What I would want Andres to do differently is to create maybe say eight different ecosystems, all separated from each other where there is no interbreeding between individuals in different ecosystems. And then set up the following parameters:

Some species age and sexually reproduce at the local level, while they may have relatives in other ecosystems that do not age and reproduce only asexually. This is not a big stretch as there are examples of this in the real world. For example, the opossums that live on a predator free island have evolved life spans that are 50% longer than mainland opossums, so they can be seen as a non-aging example for our purposes. We have seen some whiptail lizards with no need of a male, as well as the Brahminy blind snake that has colonized 6 continents-both totally asexual reproducers. So let us populate each ecosystem with aging/sexual and non-aging asexual species, and introduce an evolving predator into each one and see what happens. We will also allow occasional migration of species from one ecosystem to another on a rare periodic basis. I am guessing that if you plug all these parameters into a math model that it would take no time at all to prove that evolving predation selects for aging and sex and declining fertility and menopause almost everywhere almost all the time. This would make a great project for someone's PhD!

Now before we move on, let's consider the opposite situation, where evolution selects against aging, and against sex and against declining fertility. Just think of small tortoises that can live 120 years and more and have increasing egg output every year they get older (tortoises still have sex however (I assume) -if Brahminy Blind Snakes or whiptail lizards were more studied we might find the same situation with them - negligible senescence and increasing fertility with age.)

So in the case of the tortoise, she has solved the problem of the evolving predator for good. She has full body armor. The only things that will kill her are non-evolving forces of mortality like famine, drought, fire, cold, etc.. (I am ignoring parasites and viruses here because there is no advantage to these life forms in killing the tortoise). So if what kills the tortoise is always the same, there should be **one** particular optimal genetic combination and body plan in the (evolutionary) short run that can best defend against these forces of mortality. In the case of the tortoise in the wild, the age distribution of the population will be quite different than the youth skewed distribution we saw in Medawar's test tubes, and in prey populations subject to predation.

The age distribution of tortoises eventually will have an opposite skew where there are more older individuals than younger ones. And this happens after population crashes caused by famines droughts or fires etc. Those tortoises that survived a bunch of prior famine and droughts are the most likely to survive the new ones. So after a population crash you have all the older less diverse tortoises reproducing when good times return to repopulate the ecosystem. And it is during these times of expansion of the genes of the older less diverse individuals in the gene pool where selection against aging and against declining fertility against sex and against diversity occurs. **Most forces of mortality eventually select FOR <u>asexual</u> reproduction and FOR immortality, all of them except evolving predation.**

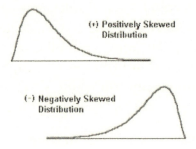

The left humped + distribution represents the youth dominated age distribution of prey species being killed by predators

The right humped distribution represents predator-defended species after a population crash/caused by famine or drought and dominated by older individuals.

Now that I look at the two distributions above, it seems possible that the age distribution pattern of the population also leads to the individual's lifetime reproductive patterns in the number of offspring produced at what age. Declining fertility for predator dominated populations and increasing fertility for predator immune populations.

Just a quick search for distribution of various animals shows that smaller groups often end up isolated from other small groups and larger populations. This provides fertile ground for the selection for sex and aging at one location and eventual migration to other areas where the same asexual non-aging species has gone extinct.

Distribution of Wild Pigs

Distribution of Black Bears

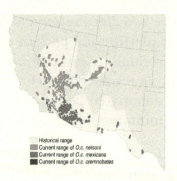

Distribution of Desert Bighorns

Chapter12: A World Without Men-The Future Course Of Evolution?

It is hard to predict how long it would take, but if this new BIG PICTURE view of evolution is correct, there could come an eventful day in human history when the first female gives birth to a genetically identical daughter with no need for a father-a true virgin birth!

When this time comes, it could be the beginning of the end of the male sex in human beings. It would likely come at a time when humans have evolved a much longer lifespan bordering on immortality! Who knows, maybe 1,000 years or even more? At this same time it will be the final vindication of Darwin's and Richard Dawkin's selfish gene view as the main driver of evolution. But until that day comes, the Selfish Gene theory needs to be taken down a notch and reconsidered as more of a subset of the larger BIG PICTURE of evolution which for now is more driven by the UNSELFISH GENEOME than the Selfish Gene, at least in the most important areas of evolution regarding aging and sex.

If left unchecked by evolving predation, the evolutionary direction for ALL organisms eventually would be a world full of identical female clones, a world full of Darwinian female monsters. It would necessarily be a safe world plagued only by the occasional mass famine, drought, or meteor strike to control population growth. It would probably be a boring world. And for those of us who are studying the aging process with the intent of prolonging our lives quite dramatically-we better be careful what we wish for!

One thing that I now suspect after having done this theoretical exercise is that asexual reproduction might be much more widespread on the planet than is currently known. I suggest that it might be fruitful if future biologists spend more time evaluating the female offspring of various animals to look for more "virgin births".

Based on the rapid progress in aging research going on right now as I write, it is more than likely that the first human beings that will live longer than 800 years or so are already alive on this planet. I once saw a statistic that if all aging and disease were a thing of the past, the average lifespan would increase to about 800 years before one would likely be killed by an inevitable accident.

If we ever did achieve the potential for 800 year life spans or longer I can just imagine how much more safety would be demanded in the world. It would be one thing to risk losing the remaining 30 aging years of your life by boarding a plane when you were age 50, than to be risking an expected 750 non-aging youthful years if there was a crash. Living so long might reduce the world's appetite for risk, again making it a more boring place to live. Again, let's be careful what we wish for, finding a cure for aging, might require finding a cure for sex types!

Bonus Chapter #13- The Evolution of Human Religion & Religious Suicide

I don't have enough material for another book right now on the topic of the evolution of human religion. So I thought I would add a bonus chapter for anyone interested in some fascinating insights one can make while viewing religion from the BIG PICTURE and the UNselfish geneOME perspective.

Of course a set of genes that motivates a person to commit suicide for the benefit of his or her religion could not possibly exist in the selfish gene view of evolution.

The closest the mainstream theorists have come to explaining sacrificing one's future potential offspring for the good of other individuals is to come up with the idea of inclusive fitness, and its more narrow version called kin selection. Kin selection can be quickly grasped by considering the biologist J.B.S. Haldane's famous quote:

"I would lay down my life for two brothers or eight cousins". Haldane's remark alluded to the fact that if an individual loses its life to save two siblings, four nephews, or eight cousins, it is a "fair deal" in evolutionary terms, as siblings are on average 50% identical by descent, nephews 25%, and cousins 12.5%. Thus, if you die, 100% of your genes die with you. If your bother dies, only 50% of your genes which he shares with you die. So from a selfish gene perspective you would not be willing to die to save a single brother.

But maybe to save two of them! Haldane also joked that he would truly die only to save more than a single identical twin of his or more than two full siblings.

So this is all fine and well when examining selfish gene theory within cohesive families, but it completely falls apart when trying to explain one's willingness to die for a religion or a country (where the selfish gene no longer plays a role). This drive to sacrifice one's self for the good of the group usually is seen in males as would be expected from the BIG PICTURE point of view as male sex types being defenses to evolving predation.

I left religious suicide out of the main text of the UNselfish geneOME book and included it as a bonus chapter because I believe the initial evolution of religion and religious suicide was not a response to evolving predation, but rather as a group defense to famine and drought. Let's see how this might work:

Let us imagine the world was populated long ago by many small groups of hunter-gatherer humans. These small bands of humans might have roamed and defended (from other groups of humans) a small territory for many generations. These groups preceded farming societies and represented a mid- point between being 100% nomadic hunters, and a settled agrarian society. Let's assume these groups were spread out all over the world. Things could go on forever like this as long as the weather stayed the same. However, eventually there would come a long drought that would eventually become a drawn out famine.

None of these hunter gatherer groups would have much in the way of emergency reserves of foods, as they had not invented farming yet. So, droughts and famines would have wiped out almost all but the luckiest of these little groups from time to time. What would have been the characteristics of these human groups that could survive famine? They would either be willing to raid and kill and eat other groups of humans, or if they were too far away, they would engage in killing some members of their own group and eating them to get through the famine.

After some of these groups later evolved into more stationary groups tied to the land due the invention and engaging of farming which produced surplus reserves for lean times, the raiding of other human groups and eating them- option became less viable. These societies would have had to depend almost strictly on human sacrifice and cannibalism to survive a drought or famine.

So the first requirement for any small human hunter-gatherer group to evolve into a longer term sustainable human civilization was for the group to embrace human sacrifice and cannibalism in times of drought and famine.

In fact the Aztecs refer to sacrificial blood of humans as "our most precious water". It starts to make sense when we see that Jesus is symbolically cannibalized at the last supper where the apostles drink his blood (wine) and eat his body (bread).

> WHILE THEY WERE EATING, JESUS TOOK A PIECE OF BREAD, GAVE A PRAYER OF THANKS, BROKE IT, AND GAVE IT TO HIS DISCIPLES. "TAKE IT," HE SAID, "THIS IS MY BODY."
>
> THEN HE TOOK A CUP, GAVE THANKS TO GOD, AND HANDED IT TO THEM; AND THEY ALL DRANK FROM IT. JESUS SAID, "THIS IS MY BLOOD WHICH IS POURED OUT FOR MANY, MY BLOOD WHICH SEALS GOD'S COVENANT.
>
> MARK 14:22-24 (GNT)

So in the beginning, when a few smart members of various small groups of humans realized that they had to engage in murder and cannibalism to survive the drought, it was probably not an easy sale to the targeted victim who likely would fight to the death to save himself. With everyone fighting to avoid being the cannibalized, the groups' extinction during a famine would likely come quickly.

Somehow a few of these groups did something differently and managed to survive. What would be the traits of the surviving groups? They likely invented the system of social hierarchy where a small group of elite humans (the king and queen and or royalty) would have priority to resources over everyone else. Thus if the famine or drought were a long one, at least the king and queen would survive to repopulate the group after the famine and drought abated. "Long live the King/Queen" now seems to make a bit more sense!

These surviving groups of humans who had evolved social structure and kings and queens (kind of like a colony of naked mole rats), when encountering a famine, would somehow stay organized even as the lower caste members starved to death while the king and queen ate their fill.

So if these were the only groups that would suvive and replace the other groups that went extinct during the famine and drought, there would then come another round of selection. There were groups where there was a king and queen, and a large number of subjects who would voluntarily sacrifice themselves to keep the king and queen alive. These groups then outcompeted the groups that simply had a king and queen with less fanatic subjects who tended to go extinct at a faster rate during famine and drought.

Now here is where the first religions likely evolved. The kings and queens knew that they were nothing special other than being born into the right place at the right time, so they had to convince the lower memebers of society that there was something special about them that was so special that they should willingly sacrifice themselves for their king or queen. The followers of the king and queen would be convinced to willingly commit suicide for them if it would help. The king and queen were likely presented to the masses as human Gods on earth, and the brainwashing of the religious followers would begin.

There necessarily would have beeen two religions in each society: the first would be the religion for the masses of followers who would be called on to sacrifice themselves in an emergency. These are the ones that in normal times would support the religion, the king and queen, and the royals.

The second religion would be a secret religion of the few ruling elite that would likely involve ways to extract surpluses from the masses for the benefit of the elite (i.e. offerings, taxes, tithes), ways to keep them amazed (miracles), and how to convince them to commit suicide if they time came. And of course the secrete religion would include various recipes for how to cook up some delicious cannibal feasts.

Those groups where the brainwashing went smoothly and generated large percentages of subjects willing to commit suidicide for God or in this case the King or Queen, became a very formidable force in the face of famine and drought and would be almost impossible to exterminate by evolution. These little groups of humans became the survving cockroaches that evoluton just could not get rid of with a famine or drought! In fact the beginning colony of the founding of America, Jamestown, is now known to have engaged in cannibalism to survive a famine, and had they not, they would have gone extinct and history may very likely have been changed dramatically.

So is there some sort of gene that has been passed down through generations that inspires the carriers to commit suicide for their religious leaders? The religious gene? I believe there is and it surfaces in various peoples at various times in their lives. There are quite a few recent examples of this phenomenon:

How about modern day religious suicide cults?

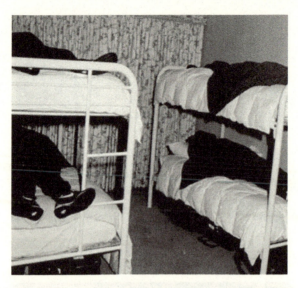

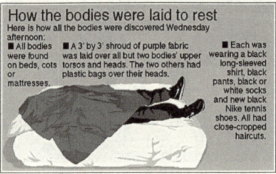

The Heaven's Gate Cult members all committed suicide at the request of their spirtual leader (God) Martin Applewhite in order for their souls to be beamed up to an alien spaceship that was hiding behind the comet Hale Bopp. Like most religious cults there was a requirement for regular <u>fasting</u>.

Origin of the Term "To Drink The Kool Aid"

Jones' brainwashing and abuse escalated to new levels as the members arrived in Jonestown. There is great evidence of mind control tactics, most notably sleep deprivation and exhaustion. The residents of Jonestown worked in the fields six days a week from 8AM until 6 PM , and after their small dinners, participated in evening agricultural meetings that would sometimes last until 2 AM . Though the residents were kept on this strict schedule, there were no calendars in Jonestown. Jones' voice would be booming over the loudspeakers without stopping, every day, all day . "There was fatigue"; "everyone was defeated and tired" . Though the residents were exhausted and wanted to sleep, falling asleep during their all night agricultural meetings was not an option. Jones told them that sleeping "proved that your head was in the wrong place, which made you more susceptible to committing treason" and was reason to be punished.

Interestingly Jones and the few leaders with him in the upper echelon of the People's Temple used to keep the followers amazed by performing various "miracles" like guessing newcomers' social security numbers, or pulling tumors out of their bodies with just their hands. In reality, they had placed some private detectives in the parking lot and were able to gain information about the new members from their license plates and other forms of information. The "tumors" removed were actually chicken parts hidden in their palms.

Movement for the Restoration of the Ten Commandments of God, Uganda

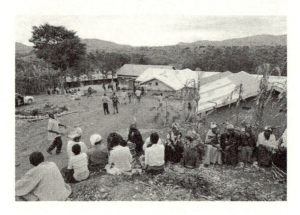

The MRTC were an apocalyptic Catholic offshoot established in the 1980s after an alleged vision of the Virgin Mary, ordering strict obedience to the Ten Commandments. The sect members spoke very little and sometimes adopted sign language to avoid bearing false witness to their neighbor, they prohibited sex to avoid adultery, **and they implemented bi-weekly fasting.** As the supposed year of the apocalypse drew near, daily confession was encouraged, the sell-off of possessions was enforced, and work in the fields ceased. However, when 'Judgment Day' failed to occur the followers began to question their leaders' authenticity, and so a second doomsday was announced for March 17th, whereby all the 1000 followers, adults and children were invited to celebrate their imminent salvation. Little did they know this would culminate in self-immolation and poisoning.

Modern day suicidal religious cults generally requried their members to fast. Why is this important? Because fasting apparently tricks evoluton into thinking that a famine is occurring which apparently triggers the expression of the religious gene which makes the carriers of the gene much more agreeable to the idea of commiting suicide for the benefit of the king or queen. Caloric restriction is well known to activate genes that both decrease the aging rate and inhibit fertility, so of course there could be additional effects triggered by fasting.

Do more ancient religions use the practice of fasting to trigger the religious suicidal impulse?

Islam- A month of daily fasting during the annual tradition of Ramadan

Christianity- Lent: the 40 days of partial fasting preceding Easter used to be much more strict in the past.

Judaism-Strict Jews do a complete fast abstaining from food and water for 6 full days per year.

Hinduism- Is full of various fasting requirements all year long

Buddhism- Fasting is so prevalent for the orthodox Buddhists that some extreme monks have gone so far as to starve themselves to death in their search for Nirvana.

And if you just Google the term "fasting" you immediately reach the Wikipedia entry which includes these additional descriptions of religions who incorporate fasting into their laws.

Bahá'í Faith
Nineteen Day Fast

Buddhism

The Buddha emaciated after undergoing severe ascetic practices.Gandhara, 2 – 3rd century CE. British Museum.

Buddhist monks and nuns following the Vinaya rules commonly do not eat each day after the noon meal.[1]

Anglicanism

The Book of Common Prayer prescribes certain days as days for fasting and abstinence, "consisting of the 40 days of Lent, the ember days, the three rogation days (the Monday to Wednesday following the Sunday after Ascension Day), and all Fridays in the year (except Christmas, if it falls on a Friday)

Eastern Orthodoxy

For Eastern Orthodox Christians, fasting is an important spiritual discipline

Methodism

In Methodism, fasting is considered one of the Works of Piety.[31] Historically, Methodist clergy are required to fast on Wednesdays, in remembrance of the betrayal of Christ, and on Fridays, in remembrance of His crucifixion and death

Oriental Orthodox Churches

All Oriental Orthodox Churches practice fasting however the rules of each Church differ. All Churches require fasting before one receives Holy Communion. All Churches practice fasting on most Wednesday and Fridays throughout the year as well as observing many other days

The Ethiopian Orthodox Tewahedo Church has an especially rigorous fasting calendar

Church of the East

The Assyrian Church of the East practices fasting during Lent, the seven weeks prior to Easter

In general, fasting remains optional in most Protestant groups and is less popular than among other Christian denominations.[36]

Mormonism

See also=Fast Sunday

For members of The Church of Jesus Christ of Latter-day Saints, fasting is total abstinence from food and drink accompanied by prayer. Members are encouraged to fast on the first Sunday of each month, designated as Fast Sunday. During Fast Sunday, members fast for two consecutive meals for a total of 24 hours.

Jainism

Fasting in Jainism

There are many types of fasting in Jainism. One is called Chauvihar Upwas, in which no food or water may be consumed until sunrise the next day. Another is called Tivihar Upwas, in which no food may be consumed, but boiled water is allowed.

Almost every religion you look into has fasting requirements, and now we know the likely reason why. It keeps the true believers believing. It sends the genetic signal to the followers to go into a mental state where extreme acts of sacrifice are possible for the group.

Now before we end this analysis, we need to notice the difference between the two kinds of suicide seen in the two categories of religions.

The first type of religion is a religion that evolved from groups that primarily depended on agriculture and farming over the eons of time. This includes the majority of religions on the planet which are religions of farming communities that have survived over vast amounts of time. The religions might change and the land farmed occasionally changed but the methods were the same, whenever there was a famine or drought, some of the members of the group were cannibalized to allow the survival of the rest. So these types of religions are the ones that from time to time can trigger the mass suicide response to various stressors, usually starvation or drought. One just needs to look at the example of Jesus, feeding the masses by magic, voluntarily sacrificing himself on the cross to make up for the sins of man as a perfect personification of the true purpose of this kind of religion.

There is a second type of religion that is more rare. It would be religion that has evolved from societies that are not primarily farmers, but primarily nomads or hunter gatherers. Religions that evolve from these types of civilizations would tend to have a suicide impulse that is not a form of self-sacrifice for the good of the group, but rather a suicidal impulse to attack other nearby groups in order that their "home" group could swoop in and claim the resources of the defeated group and thus ride out the famine.

The main character worshipped by this type of religion would more likely be some type of warrior.

Eventually religions of the second type should go extinct, as once they have vanquished all the other nomadic groups in an area, they will ultimately have to be attacking religious groups that evolved from farming civilizations. And farmers, being tied to the land and their homes, are likely more determined and resourceful than nomads.

In the modern age, the ancient evolutionary adaptation of religious suicide keeps rearing its ugly head. We will likely continue to be bedeviled by this shocking primitve instinct in its various manifestations for many generations into the future until the religious suicide gene eventually becomes extinct from the human gene pool, or until the world becomes so stress free that it is never triggered.

Bonus Chapter #14 What chemicals and molecules that are good for you have in common-

As I type, I am expecting that this chapter is going to be the one I enjoy writing the most. Why? Because I am going to present to you what seems like very complicated information about the various chemical structures of substances that are known to be very good for you and that can even extend your lifespan and retard aging. And when you are finished with this chapter, I hope and expect, that you will get that big "AHA!" feeling, and it will all seem quite simple. I also hope that when you are finished reading this chapter that you will be able to look at the chemical structure of almost any substance and be able to have a good idea if it would be good for you or not.

Let me start with my favorite example: an herb well known amongst those who practice Chinese medicine called Fo-Ti Root. Its Chinese name is *He Shou Wu* which roughly translates as Mr. Wu's hair stays black. One story about how the Chinese name (He Shou Wu) for this herb came about tells of a General who was convicted of a serious crime and sentenced to death by confinement to a remote cell that was dug into the ground with no access to food or water. After a year, upon returning to the cell to have his remains removed, his executioners were surprised to find that not only had General He survived, but he had gone through a complete rejuvenation that had been able to reverse gray hair on his head back to black. It turns out he had survived exclusively on a vine that grew in the crevices in his cell walls... the Fo-Ti root. (Notice there is some caloric restriction going on here too!)

Another version of the story is that at 58 years old, Ho Shou Wu could not conceive a child. After being advised to take Fo-Ti Root for his problem, he fathered many children and amazingly his hair turned from gray to black while his bo[dy] amazingly looked years younger. He supposedly lived to 160 years old with his children living to 130 years old. All attributed to them taking Fo-Ti Root.

So let's take a look at the chemical structure of Fo-Ti Root. This is easy, all you need to do is search Google using the search terms Fo-Ti Root and scientific name and you get Polygonum Multiflorum.

The next step is to search google IMAGES for the terms Polygonum Multiflorum and chemical structure and you will find the following chemical diagram:

Fo-Ti Root Chemical Structure

If you are new to this, this chemical structure will mean nothing to you, so I will cheat for you: I instantly recognize within this chemical is a molecule of resveratrol attached to some other chemicals that can easily be snipped off.

Now below check out the structure of the modern miracle comound shown to extend life span in all sorts of organisms: resveratrol!:

Resveratrol seemed so promising when discovered that some Harvard researchers formed a company, Sirtris Paharmaceuticals to commercialize it as a drug, after they tweaked its structure to make it patentable (they claim they wanted to to make it more effective).

So while humans studies may be ongoing right now for how resveratrol and its variants (analoges) can affect people over the long term, we will have to wait many decades for the final conclusive results. However if we want to take a sneak peak at how long term resveratrol supplementation will affect human health and longevity, all we need to really do is dig into the history and reports fo Chinese medicine from long ago! An excellent web page to begin looking into the long history of Fo-Ti Root can be found at :

http://www.itmonline.org/arts/hoshouwu.htm

and after a quick persual I noticed that legend has it that one should only take this root (and probably resveratrol) every other day, and to stop taking it for awhile if you get the "sweats".

Now here comes the fun part. We will now have a little lesson about how organic chemical structures are depicted. It is very simple once you get the hang of it.

Organic chemistry deals mainly with four chemicals, Hydrogen (H), Oxygen (O), Nitrogen (N) and Carbon (C) and from time to time you get some Phosphorus (P). Other chemicals are seen from time to time, but these 5 chemicals will take care of most of the organic molecules.

I chose the order of the molecules below on purpose, as you see H only has one binding site that other chemicals can attach to. O has two attachment sites, N has 3, C has 4 and P has 5. So if you were looking at these chemicals like a lego set they would look like this:

$$H- \quad -O- \quad -N= \quad =C= \quad =\underline{P}=$$

And where a line is coming out of the molecule you can attach any line from any other molecule to create a bond. So if one hydrogen bonded to another hydrogen it would look like this
H—H

and would be known as H2.

Everyone knows water is H2O and it would look like this

H--O—H.

A nitrogen and three hydrogens would bond to become an ammonia atom that looks like this:

and if you bound a carbon with 4 hydrogens you would get methane , the flammable gas given off by vegetation eating mammals:

$$\begin{array}{c} H \\ | \\ H - C - H \\ | \\ H \end{array}$$

So this is the very simple concept of organic chemical bonding. And it doesn't have to always be hydrogen, all the chemicals shown can bind to all the others at their binding sites in any way imaginable.

Now what I am about to show you is how to tell, just by looking at a chemical structure, if there is a good chance it can improve your health and extend your lifespan.

Let's start by looking at a "good" steroid hormone that decreases with age, let's start with pregnenolone:

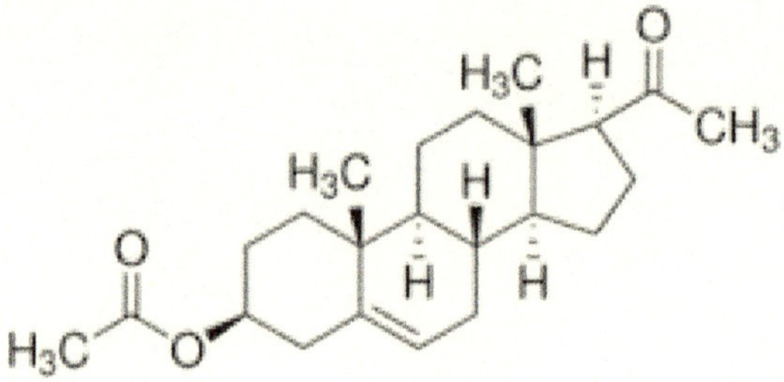

What you need to know here is that all the rings are made by carbons bonding to each other, but they don't write down all the C's as it would be too time consuming. So the rings above if they were drawn out completely would look like this:

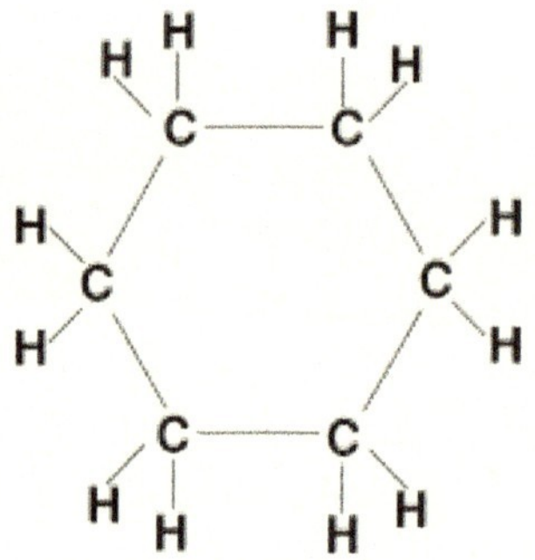

So there is a bit more to this but we know enough to now describe the three principles of a molecule that suggest it might be good for you.

1. If a molecule has an oxygen double bond dangling off the molecule that looks like this **O=** it can act as an antioxidant. You can also get an antioxidant effect of there is an **–OH** dangling off of it.

2. If there is a **–CH3** dangling off of the molecule it can act as a methyl donor. Remember you need methyl groups attached to your DNA to suppress aging genes. And the amount of methylation left on your DNA can tell you how long you have left to live.

3. If the molecule has something close to the shape of a steroid hormone (like pregnenolone above) or DHEA or progesterone, or testosterone, it also suggests the molecule might be good for you as this strucure allows it to bind to DNA.

Why do steroid hormone structures allow the molecule to bind to or associate with the DNA?

As most people know, your DNA is made up of 4 chemicals G, C, A and T. Their chemcial names are guanine, cytosone, adenine and thymine. They basically act as letters in an alphabet.

What is important here is that DNA consists of 2 strands, and the G on one strand always binds to a C on the other strand, and the A on one strand always binds to the T on the other strand.

Here are what the 4 nuucleotides G,C A and T look like as chemical structures:

adenine (A)   guanine (G)   cytosine (C)   thymine (T)

Now when a G binds a C or an A binds a T in DNA it looks like this: (does it seem familiar?)

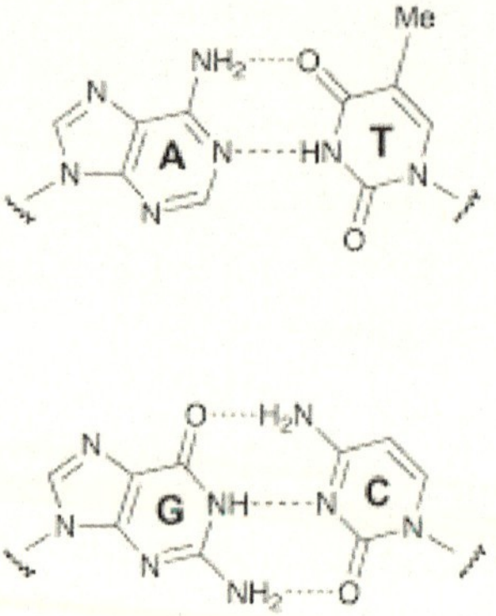

Above are what DNA base pairs look like when paired in the DNA

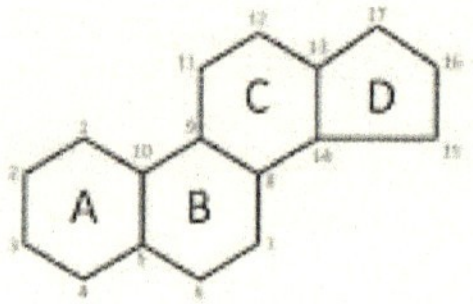

Steroid skeleton

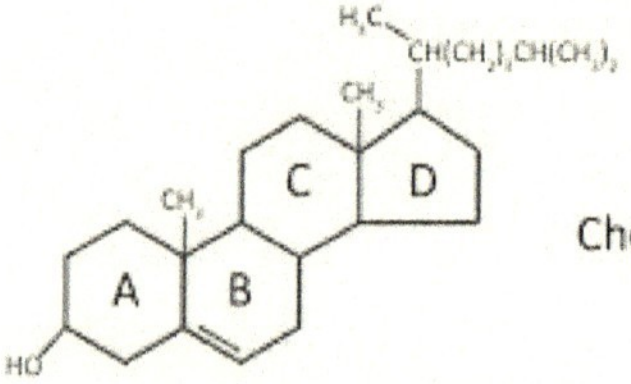

Cholesterol

See how the shape of DNA base pairs are very similar to the steroid hormone skeleton above?

Here are some steroid hormones below, they are all structurally similar:

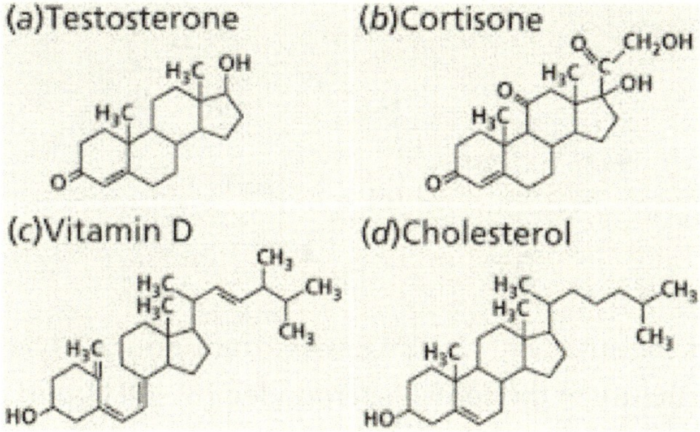

(a) Testosterone (b) Cortisone (c) Vitamin D (d) Cholesterol

Now notice that they all have CH3's attached to them, as well as =O's and –OH's, these all have the three elements of things that are good for you- a steroid structure, methyl groups (–CH3's) attached, and antioxidant capabilities =O , -OH.

Now let's take another look at resveratrol:

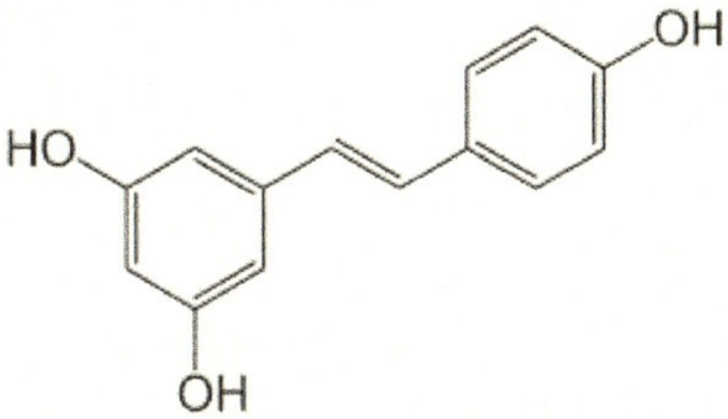

It seems to have a structure that would be compatible to binding with your DNA, and it is also adorned with 3 –OH's which have antioxidant potential.

Now let us take a look at the active chemical in a plant known a astragalus, which is one of the few componds tested shown to cause your telomeres to lengthen. In fact, scientists have clipped the active compound out of astralagus and now market it as an expensive anti-aging supplement called TA-65. (Astralgus is much cheaper).

Above is the structure of astragalus extract, notice it has it all: a steroid structure in the center, surrounded by –OH's and –CH3's!

TA-65

To make TA-65, scientists simply had to snip off two of the dangling carbon rings of astragalus.(Note that in these strtuctures if there is a line sticking off with nothing at the end, it is assumed there is a CH3 attached-they do this for convenience).

With our three little rules we can now examine various natural substances that have long been rumored to have miraculous medicinal properties. Let's look at ginseng first:

Look how very simialr it is to TA-65 above

Ginseng

Another famous Chinese herb which is rumored to have been responsible for the extreme longevity of a small group of Chinese farmers who regularly drank it as tea is known as Jiaogulan Tea. Again it has many similarities to atralagus and ginseng-A steroid structure surrounded by many antioxidants and methyl groups.

Jiaogulan Tea Chemical Structure

Now you don't always have to have a steroid structure, the following is a molecule called Buthylated Hydroxy Toulene (BHT), a well known food preservative said to be an antioxidant. It turns out that BHT is one of the few compounds tested that can increase both the average and MAXIMUM lifespan of the mouse. To me BHT seems to be more of a methyl donor than an antioxidant. I have heard it is also good for reducing hangovers.

BHT

Many cultures have a place for green tea in traditional medicine and many modern studies have noted many health benefits of regular green tea drinking. So let's look at the structure of its active ingredient EGCG:

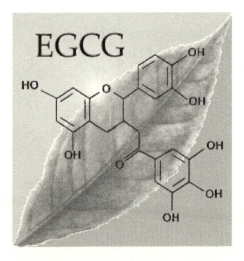

So we can see in the active ingredient of green tea (EGCG) a hint of steroid structure with lots of antioxidant attachments.

So you pretty much have it down now, and you can go to google images if you are curious and take a look at various drugs as well as hormones and supplements .

Let's take a look at ibuprofen for example:

Ibuprofen looks like it should be good for you! It has the shadow of a steroid structure and is loaded with antioxidant groups and methyls to donate!

So you could then look up ibuprofen and lifespan on google and look what pops up:

VIOLA!

Ibuprofen, a common over-the-counter drug worldwide, added to the healthy lifespan of yeast, worms and flies in a recent study.

Credit: Texas A&M AgriLife Research/Kathleen Phillips
A common over-the-counter drug that tackles pain and fever may also hold keys to a longer, healthier life, according to a Texas A&M AgriLife Research scientist.

Regular doses of ibuprofen extended the lifespan of multiple species, according to research published in the journal *Public Library of Science, Genetics*.

"We first used baker's yeast, which is an established aging model, and noticed that the yeast treated with ibuprofen lived longer," said Dr. Michael Polymenis, an AgriLife Research biochemist in College Station. "Then we tried the same process with worms and flies and saw the same extended lifespan. Plus, these organisms not only lived longer, but also appeared healthy."

He said the treatment, given at doses comparable to the recommended human dose, added about 15 percent more to the species lives. In humans, that would be equivalent to another dozen or so years of healthy living.

I will not add any more dialogue but just give you a few more examples of various things that are good for you so you can ponder them on your own:

Ascorbic Acid (Vitamin C)

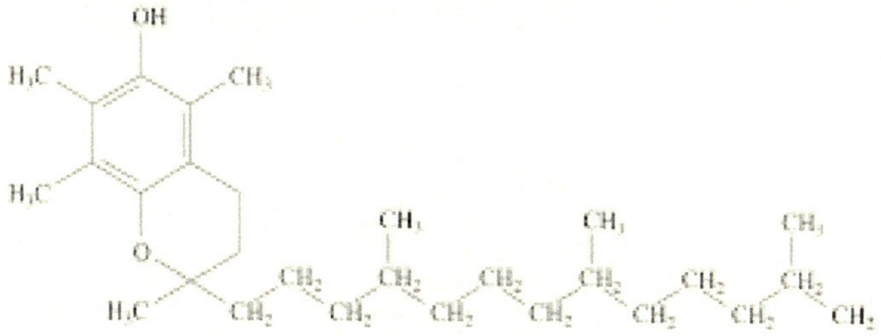

Vitamin E (α-tocopherol)

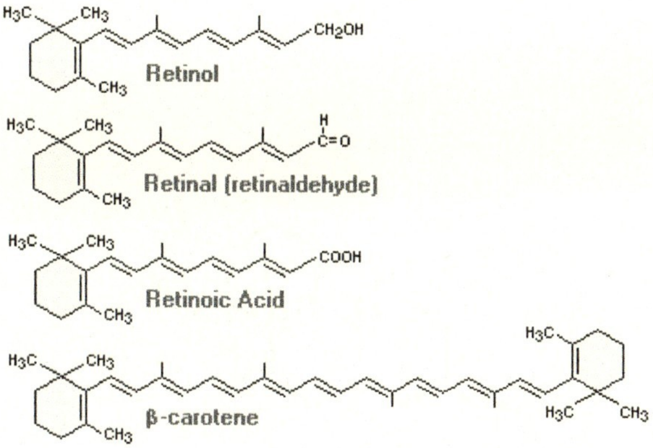

Various forms of Vitamin A

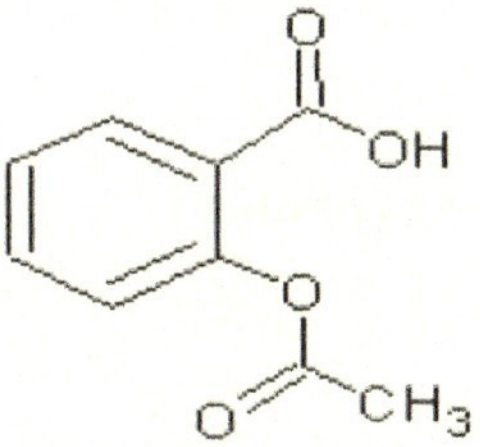

- Retinol
- Retinal (retinaldehyde)
- Retinoic Acid
- β-carotene

Aspirin
Acetylsalicylic Acid
$C_9H_8O_4$

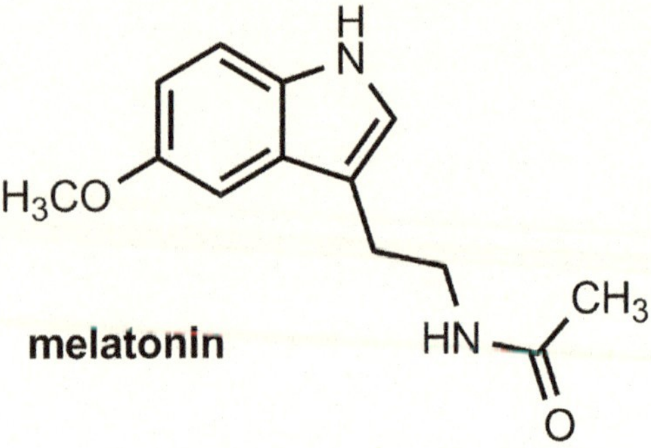

melatonin

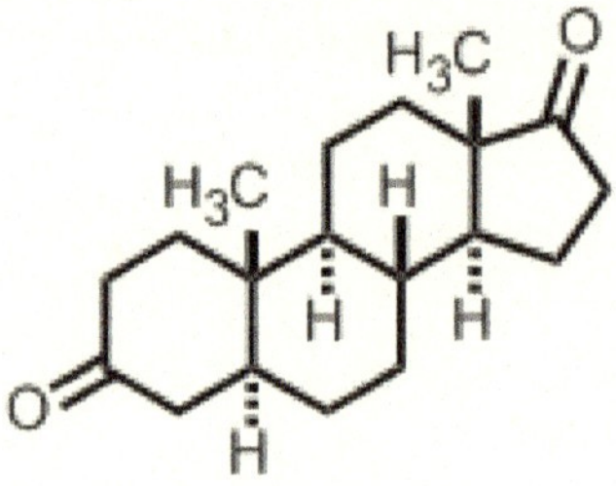

DHEA

Progesterone

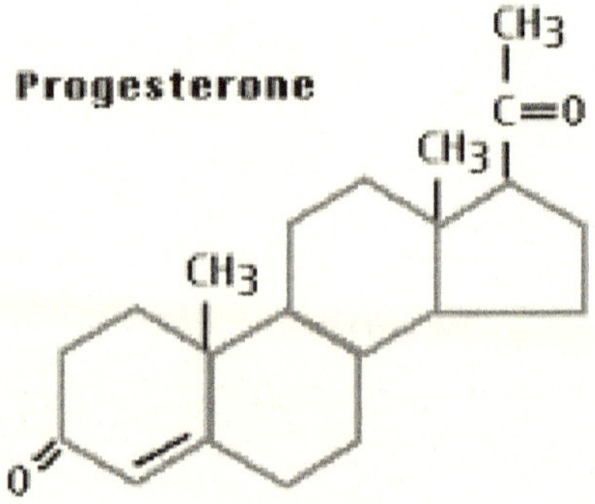

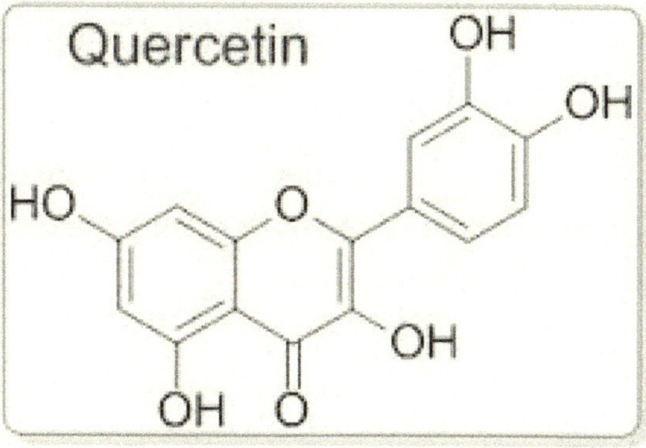

Quercetin

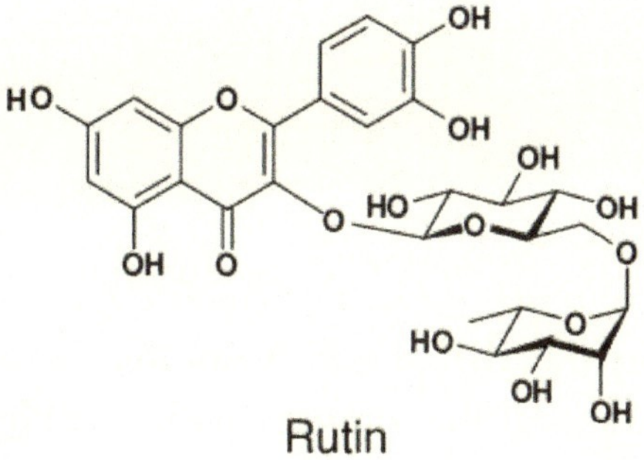

Rutin

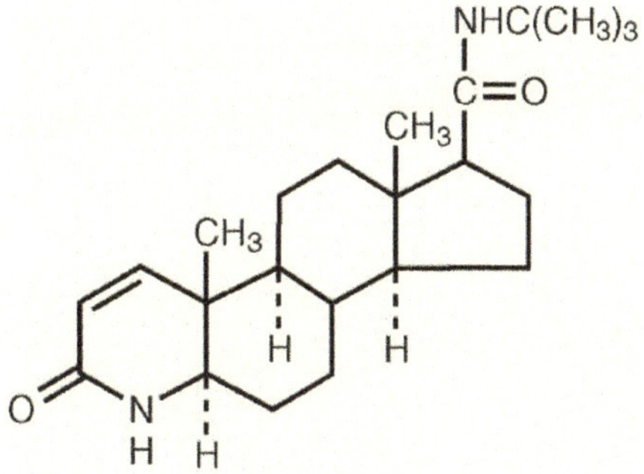

Proscar-

Prescribed for Prostate issues and regrows hair (is almost the same as progesterone)

I could go on and on, but you get the point. So have some fun from time to time and look at chemical strcutures of various things that are supposed to be good for you be it a drug, a hormone or a supplement.

But wait! Now there is one more category of substances that are really good for you that seem to have their own special rule which I haven't completely figured out yet.

The rule is that they look like they have most of the chemcial building blocks to make a guanine

[Structure of guanine]

Guanine (G)

Now you could say that this provides half the steroid structure, but I believe that compounds that mimic guanine just by itself or can be converted into it seem to be very potent lifespan extenders.

Metformin:

[Structure of metformin]

Deprynyl (generic name Selegiline)

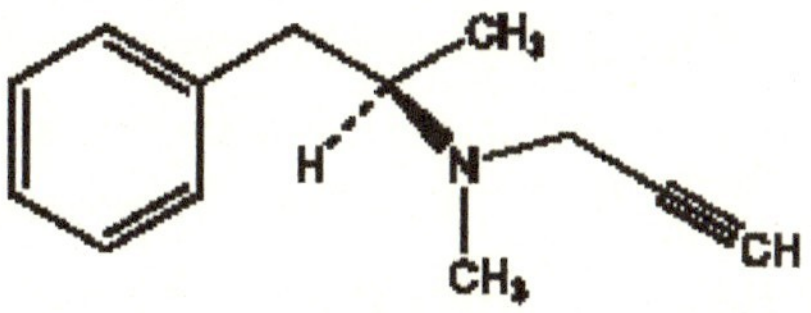

Selegiline Base

And a new life span enhancer was just discovered- a compound found in soap!

Allantoin

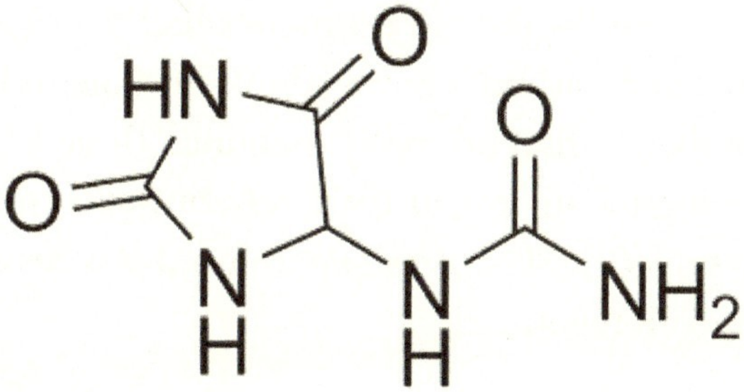

This fourth rule as you can see is a bit murky and a bit sloppy for now, so if any of you can figure out a better rule that can explain the actions of these molecules please let me know.

My contact info is Jeffbo at aol dot com.

Now if you run these ideas by your traditional scientist schooled in hormones and hormone receptors they will be quite skeptical I expect. Why? Because the traditional thought is that each hormone fits into a specific receptor and generally has no other function than to act as a key that fits into a lock. I believe this is only half true for many hormones and substances. For example if you look at Vitamin D3, there are receptors for Vitamin D3 that when D3 binds to it, the hormone-receptor complex binds DNA to control various genes. All well and good.

I do however expect that various hormones, and especialy the steroids, will also be found to associate directly with the DNA without the need for a receptor. Call it a secondary function that gradually controls aging by controlling the methylation content of DNA. There are quite a few drugs that do not need receptors to function and directly bind to DNA, so this idea is not that speculative.

In my first published paper from 1998 I hypothesized that the "good" hormones were involved in methylating DNA. I had assumed that somehow they were involved with donating methyl groups to the DNA and this was catalyzed by antioxidants. I did not know at the time that the steroid hormones had the exact shape of DNA base pairs, and had both antioxidants and methyl groups dangling off of them. I had orignally imagined three separate substances coming together near the DNA, but what we have found with this analysis is that the three separate substances are often found connected all together!

There are a few other substances that can extend lifespan that I have no idea how categorize as they are quite big and seem to follow their own rules. For example take a look at Rapamycin which has been shown to extend mean and maximum life span in mice:

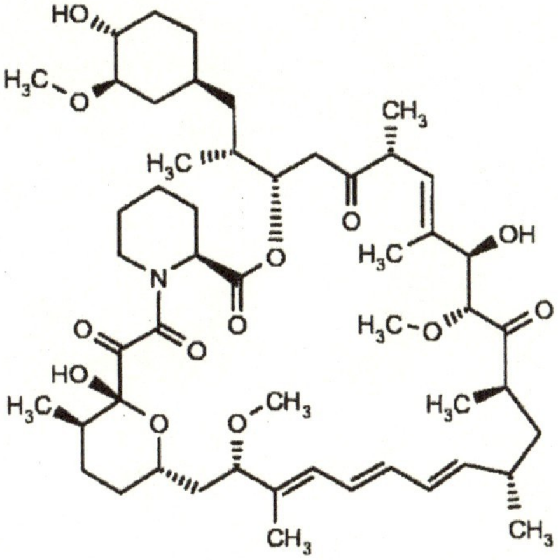

Rapamycin

I might dare to say that I see the potential for a steroid shape after Rapamycin is broken down in the body, and we definitely see plenty of antioxidants and methyl groups. So who knows maybe it does follow our rules.

So what do the hormones that increase with aging and age us have in common? At this point I can only note that bad hormones tend to be big. I have always thought that the larger the hormone, the older it is in evolutionary terms as every time a hormone became larger, there likewise had to evolve a larger hormone receptor. The diagram below of course suggests that TSH should be studied as a potential aging hormone as well.

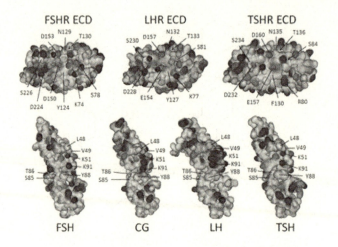

The Hormones FSH, hCG, LH and TSH

All I could find on TSH increases wih age was the following chart which suggests that the percentage of the population with elevated TSH levels tends to increase from 2% to 14% from age 19 to age 80. Also another subset of the population experiences lower TSH levels with age. Maybe TSH is an exception to the tendancy that the big hormones are bad.

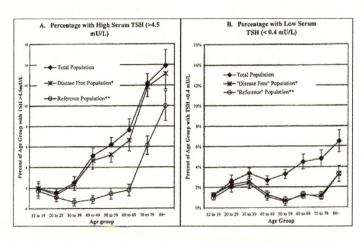

Appendix A: My Thoughts on Charles Darwin & Some Deleted Text

While editing the main text of this book, I decided to remove a few things that were making the Preface and Introduction too long. Most readers will not be interested in them, but I add them here for the few that might be:

My Research Technique:

"I began spending 10 hours a day 7 days a week in the Northwestern Med School library reading study after study regarding aging. I read the latest studies and also found myself down in the basement digging up articles and studies about aging going back to the 1920's. Some articles that seemed to contain special knowledge I would read and reread many times.

I also was using the in house computers at Northwestern to search the Internet database called Pub Med which contains summaries (abstracts) of all the science articles and studies ever conducted going back to 1967 and even further in some cases. What I discovered was that all the experiments you could ever think of had almost all been done already. So in effect for a theorist, all you had to do is ask yourself if this is true then this other thing must be true. You could then conduct an experiment to test the validity of your assumption. And rather than having to do an elaborate experiment that might take several years and many hours of time, you could get the answer to your experiment in about 10 seconds! Just type in the right search words for the experiment you want to run and 90% of the time someone else has already done the experiment!

I started to think that all the information needed to solve the riddle of aging already existed in Pub Med, it was just that no one has ever put it all together yet to answer the question. So that was what I did for years.

Like trying to put together a jigsaw puzzle, I spent years collecting puzzle pieces, and trying to figure out how they were all interconnected and related.

After Darwin put forth his theory of evolution of the survival of the fittest, it was later re-discovered that Gregor Mendel had previously discovered the unit of information that determined what aspect of an individual would be considered fit or unfit. Mendel had discovered the existence of the gene. The basic building block of what makes up an individual organism-the gene. An individual was just the expression in the flesh of the information contained in all his or her genes.

From this starting point, evolutionary theory just kept building up more and more complicated theories and levels of explanation all based upon that irreducible piece of information called the gene. This was of course a reasonable way to try and explain how evolution worked. IT WAS A BOTTOM UP APPROACH. It sought to explain everything starting with the basic building block, the gene, and working the theory up to higher and higher levels of complexity and understanding.

I approached the explanation of aging, however, in the opposite manner, from a top down approach, and luckily for me I just gathered facts about aging from wherever I could find them. I looked at aging differences between species, how aging was affected by lack of food (it is suppressed), how it was affected by reproduction, I gathered all the higher level facts and worked backwards eventually getting down to the lowest level-the gene.

By working on higher order information about aging, I was able to avoid getting sucked into the quagmire of what is the current mainstream thinking about evolution, of starting with the simplest of concepts and then trying to expand it to a higher level of explanation.

Charles Darwin-The Giant-

Now don't get me wrong. I am in no way minimizing the incredibly huge contribution of Charles Darwin and or subsequent scientists / theorists have made to the science of evolution and biology. If you study Charles Darwin's prolific writings in detail, you cannot escape the idea that he was a giant in science of his time. He wasn't just a naturalist who came up with the idea of evolution. He was one of the first people in the world to understand that the age of the Earth was much older than the 6,000 or so years that the bible implied. He needed to understand this idea before he could even begin to formulate the theory of evolution. And he gained this understanding on his famous trip on the HMS Beagle while contemplating the layers upon layers of chronological fossil formations he would see in exposed mountainsides in what are now the seacoasts of Brazil, Argentina and Chile. These contemplations prepared him for understanding the differences between similar species he saw on the islands of the Galapagos that allowed him to come up with the theory of evolution.

And yet, after seeing these species differences between almost identical species on different islands, he was initially still skeptical of the idea of evolution.

Charles Darwin also might be thought of as the person who invented an old fashioned version of the internet. Once he returned home with his ideas he embarked on a furious and almost maniacal quest to assemble facts from all over the world to either prove or deny his emerging theory. Darwin from his country estate near London, sent out countless letters and telegrams all over the world soliciting facts about the natural world from other naturalists. He was like an internet website that attracted and requested information about various species around the globe. Darwin became a clearinghouse for all information relevant to the idea of evolution , and he maintained this clearinghouse for many years until he was finally confident that his theory was correct.

Charles Darwin was in a single word: completely totally amazing a- GIANT!

Had Charles Darwin lived longer, and had more facts to work with, he would have easily figured out what I am about to share with you as the correction and completion of his theory. Unfortunately, aging and death, interrupt the work of all of us.

Darwin figured out that all life forms are subject to the law of the survival of the fittest. He noted that minor variations in the physical attributes of individuals born into this world make them more or less suited to the environment in which they exist. Those more suited to the environment survive and pass their superior attributes onto their young who become more numerous. Those with inferior attributes perish and have fewer or no offspring and thus their attributes become more rare in the population and eventually their genes become extinct.

Darwin did not know about genes as the mode of inheritance but he had the overall concept correct. Years after Darwin's death science discovered that the mode of inheritance of physical attributes was driven by the inheritance of genes by the offspring of a parent. This simple idea has driven the expansion of Darwin's theory from its beginning to today's mainstream consensus where modern evolutionary biologists believe that all of evolution is driven by the "selfish" gene that wants to make as many copies of itself as possible-no matter what!

The problem with this is that this idea is true in many cases. And here is where I diverge with modern mainstream theorists.

In many other major cases this idea is wrong. There are instances where the spread of genes are restrained- for the good of the group (actually the species) . The group is ignored by modern Darwinian theorists because they believe that evolution can only take place at the individual level for the good of the individual. (And when I say good of the individual, I and they mean for the good of the individual being able to reproduce and spread his or her genes.).

Appendix B: Ageing: Theory needs to be revised

Here is an excellent article I found while researching this book-

Ageing: Theory needs to be revised

December 13, 2013

The existing evolutionary theories of ageing need to be revised, according to a new study, which shows that many of Earth's plants and animals grow old in surprising ways.

Keywords: Ageing, Animals, Evolution, Plants

By: Lise Brix

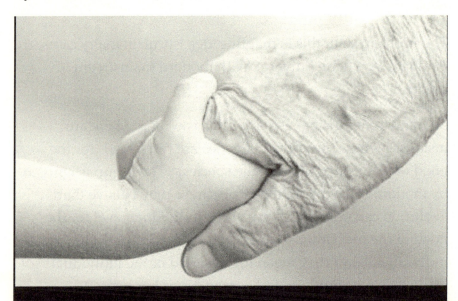

The older we humans get, the greater our risk of dying – just as the prevailing evolutionary theories of ageing state. However, if we take a look at the desert tortoise and certain marine animals and plants, we get an entirely different picture – here, the oldest members of a group have a lower risk of dying than the young members. (Photo: Shutterstock)

Everything ages – and the older we get, the greater the risk of dying.

This is the popular conception of ageing, but new research suggests that not all species age in the same way – far from it.

A Danish-led group of international scientists has studied the ageing processes of a broad group of animals, including humans, and plants, and the results indicate that the existing evolutionary theories of ageing need to be revised.

The new study demonstrates that for several species, the risk of dying decreases as they age.

Researcher: modify the theory

"According to the classical evolutionary theories of ageing, one would predict that all organisms grow weaker and have an increasing risk of dying as they grow older, just like we see in humans. Many people – including scientists – believe that this applies to all species, but that is not the case," says ecologist Owen Jones of the University of Southern Denmark, who is the lead author of the new study.

Data from 46 species

In the study, researchers used data from 46 species, including humans, baboons, birds, oak trees, lice, seaweed, crocodiles, killer whales and lions.

For each of the species, the researchers drew charts of how the mortality and fertility change across the life course of the species.

The desert tortoise (Gopherus agassizii) is one example of animals whose mortality declines with age. (Photo: Shutterstock)

Existing theories assume that these two parameters follow a fixed pattern: that mortality increases with age, while fertility decreases.

However, having compared their results across the 46 species, the researchers found that this ageing pattern is far from being a universal law of nature.

Some animals do not age

In addition to showing that some animals have a decreasing risk of dying as they age, the study also shows that other species, such as the hermit crab and the freshwater polyp hydra, have a constant mortality throughout their life.

Neither the hermit crab nor the hydra experience bodily decay as they age, and according to the researchers, this can be interpreted as if the animals actually do not age at all.

"The hydra has a constant low mortality throughout its life. In principle, this means that the hydra is biologically immortal. In nature, however, the hydra dies, for example from being eaten by predators," says Jones.

Crocodiles grow more fertile with age

When it comes to the ability to produce offspring, the researchers also found several exceptions to the existing theories.

Some species actually grow more fertile as they age. In this study, this applied to the freshwater crocodile and several plant species, such as the agave.

The alpine swift, a bird, experiences increasing fertility almost throughout its life, while sexually mature baboons (Papio cynocephalus), on the other hand, maintain a steady ability to produce offspring throughout their lives.

However, when looking at e.g. theCaenorhabditis elegans roundworm, we see an entirely different pattern. According to the study, they are actually fertile from early in life, after which they very quickly lose the ability to have offspring.

Ageing remains a mystery

According to Jones, the different patterns of mortality and fertility show that ageing remains a poorly understood phenomenon:

"In our study, we have looked at ageing in a wide variety of species from throughout the tree of life. I believe we can learn a lot about how ageing works by looking at the strange deviations from the standard, human-like, way of ageing."

Humans age quickly
One of the most surprising findings was that out of all the species featured in the study, humans age the quickest.

"As modern humans, we do not regard ourselves as organisms that age particularly quickly compared to species like mice or other animals. But out of all the species we studied, humans were the species that ages most quickly, in relative terms."

It may seem a bit odd that humans age more quickly than e.g. the fruit fly, which has a lifespan of only a few months.

We have looked at ageing in a wide variety of species from throughout the tree of life. I believe we can learn a lot about how ageing works by looking at the strange deviations from the standard, human-like, way of ageing.

Owen Jones

Jones explains this referring to the new standardised chart that his research team has developed, which enables them to compare mortality and fertility across species – despite the species having highly different average lifespans.

On this chart, humans have the steepest mortality curve of all the 46 species.

Study revolutionises our view of humans
"If, for instance, we look at the hydra, we see that its risk of dying remains constant throughout its life. For humans, on the other hand, the risk of dying increases 20-fold when we grow old, compared to the average risk of dying throughout life," he says.

This turns our conception of ourselves on its head: we do not age slowly; we actually age very quickly."

The prevailing theory of ageing
The existing evolutionary theories that explain why animals and plants age were developed back in the 1950s and 1960s.

In short, the theories state that all species, including humans, invest solely in their own survival until they are sexually mature.

After that, investment in reproduction becomes more important even at a cost of their own maintenance and survival. A consequence of these theories is that the body starts to age, or decay, from when the individual becomes fertile.

No new theories have solved the ageing mystery

This turns our conception of ourselves on its head: we do not age slowly; we actually age very quickly.

Owen Jones

The researcher says that it may not make much sense that evolution has not come up with a way of preserving the body and preventing ageing, especially considering that building an organism in the first place is such a complicated and impressive task.

"Up to now, the explanation has been that we are presented with the choice between investing our resources in reproduction here and now, or investing them in preserving our bodies and repairing damage," he says.

"There is an advantage in reproducing as quickly as possible is if you risk dying in an accident, or by predation, tomorrow. This is one of the main points in the evolutionary theories of why we age."

However, the new study shows that this prevailing theory is far from complete:

"If the theory were correct for all species, we would expect that the risk of dying would increase after sexual maturity for every species. In other words, we would see a pattern of a gradual reduction in survival probability after reaching sexual maturity for all species. But this is not the case. Far from it," says Jones.

The researchers behind the new study do not, however, have a full-fledged theory that can replace the prevailing ones. This means that the question of why most life on our planet ages and dies remains an unsolved mystery.

Appendix C: Guide: Seven Incredibly Old Mojave Desert Plants

An interesting article about very old desert plants-

Guide: Seven Incredibly Old Mojave Desert Plants

by [Chris Clarke](#)

January 20, 2012

There are some ancient plants in this photo, but not the ones you're guessing. | Chris Clarke photo

Editor's Note: *Driving between Los Angeles and Las Vegas may summon up those middle-of-nowhere feelings. On the outset, the vast spaces of monotonous desert look empty, but take a closer look with the help of this guide and you'll be spotting plants that can or have lived thousands of years.*

Chris Clarke, a California desert resident and advocate, earlier this month spoke at the California Native Plant Society's Conservation Conference about old-growth plants in the desert. He's also [a regular KCET.org commentator](#) and here shares what's easily overlooked.

The desert is a harsh place to live. Plants that grow here for more than a single growing season grow slowly, a few inches or less in a good year. And as is the case with the fabled bristlecone pine of California's White Mountains, which can live for 5,000 years or more, that slow growth habit can bring with it immense longevity. Many of the plants native to the Mojave Desert have astonishingly long lifespans. But not necessarily the plants you might guess.

It's easy to find references to "ancient Joshua trees," for instance, and people will tell you of trees with ages upwards of 700 years. The trees can certainly *look* ancient; gnarled and twisted and battered. But as it turns out, that's not the case. It's hard to determine Joshua trees' ages precisely, as their trunks lack annual rings, but based on the rate at which the trees grow, it looks like most die before their 200th birthday, with almost none reaching 300. That's impressive enough compared to our measly threescore years and ten, but it's not bristlecone-caliber ancient.

A relatively young clump of Mojave yucca, only about 500 years old or so | Chris Clarke photo

Mojave Yucca

The Joshua tree's cousin the Mojave yucca (Yucca schidigera) is a different matter. Sometimes mistaken for a Joshua tree despite its coarser build and the fact that it rarely branches, a Mojave yucca can outlive its more graceful relative by many centuries. An individual Mojave yucca plant grown from a seed will, when it reaches maturity after a century or so, grow little side shoots that eventually become full-fledged adults themselves. Those side shoots have their own side-shoots, and so does the next generation, and the next. Eventually the original shoot will die out and decay, leaving a ring of yucca stems that are, unless their subterranean connection is severed, essentially all the same plant.

We can estimate the age of a group of these clonal shoots by measuring its width and calculating how long it would have taken to reach that width. Estimates of growth rate for Mojave yucca clumps vary by about a factor of three – a foot wider each 30 years, or each 100? – but even using the more conservative 30 year rule of thumb, Mojave yucca clumps in excess of 700 years abound throughout the Mojave Desert. One ring near Lucerne Valley was described in the New Scientist as in excess of 12,000 years of age, quite possibly an overexuberant estimate. That particular ring is *certainly* several thousand years old, however.

King Clone | USFWS Photo

Creosote Bush

That's not to say there aren't 12,000-year-old plants in the Mojave, though. Take the creosote bush, *Larrea tridentata*, which possesses the same habit of forming clonal rings that expand, very slowly, over the millennia. The best-known of these creosote rings is "King Clone," near Landers, dated by biologist Frank Vasek at about 11,700 years of age. When the creosote seed from which King Clone grew hit the soil, it might have been tamped down by a mammoth or a Shasta ground sloth.

King Clone is unbelievably old, but there are plenty of creosote bushes in the Mojave that are merely astonishingly old. One specimen in a Lancaster city park has been estimated at about 800 years old. It's a nice looking plant, large and rather impressive - and there are many thousands of creosote bushes just like it throughout the Mojave.

Big galleta grass meadow in Imperial County | Chris Clarke photo

Big Galleta Grass

Even the lowly bunchgrasses in the Mojave can attain significant age. In a photography study of 19th century photographs from the Grand Canyon area, one of the plant species found to have persisted for more than a century was the unprepossessing native bunchgrass big galleta (*Pleuraphis rigida*). A lifespan in excess of 100 years isn't bad for a grass.

Big galleta is easily recognizable: it grows throughout the Mojave (and elsewhere in the California Deserts) in washes, on broad plains, in clefts in rocks, and just about anywhere else it can get a toehold. In good conditions, like those shown here in the Colorado Desert west of El Centro, a galleta clump will get to be about three feet tall and as wide. It's a favored browse plant for many animals including bighorn sheep and desert tortoise, and it serves as a nurse plant for hundreds of other desert plant species, providing shelter and camouflage as tender seedlings slowly harden to the rigors of the desert.

Buckhorn cholla | Photo by Ryan Orr/Flickr/Creative Commons License

Buckhorn cholla

Among the plants that take advantage of galleta's nursery are chollas, those fiercely armed jointed cacti with the intimidating spines. At first glance, chollas would seem like excellent candidates for serious longevity, and many of them do in fact live for centuries. The rather nondescript buckhorn cholla, *Cylindropuntia acanthocarpa,* may well outlive most of its kin.

Ranging from the easternmost reaches of the California Mojave through Nevada, Utah, and Arizona into northern Mexico, buckhorn cholla differs from many other chollas by its moderately sparse coat of spines, and the unique purple-red color of its floral filaments, a nice contrast with its (usually) yellow flowers. When it's happy a buckhorn cholla can reach 10 feet tall, but three is more usual.

To my knowledge no one has nailed down a reliable figure for buckhorn cholla longevity, but an article published in 2000 offers an intriguing hint that that longevity may be very long indeed. In a 15-year survey of a plot of land in the Sweeney Granite Mountains Desert Research Center in the Mojave National Preserve, Martin Cody and his colleagues charted the "births" and deaths of shrubs on that plot from 1981-1996, and extrapolated the likely maximum lifespans of many of the species growing there. Some species turned out to have impressive potential lifespans indeed: more than 700 years for east Mojave buckwheat, 425 for spiny *Menodora*, and a couple species with even longer lifespans. Four species had no casualties over the 15-year study period, and so the researchers could not establish a likely maximum lifespan. They were Mojave yucca, creosote bush, buckhorn cholla, and one other shrub we'll get to in a minute.

That study doesn't offer enough data to say conclusively that buckhorn cholla can live for millennia, but given the other species on the "too long to measure" list it sure looks promising. It may well be that that pesky buckhorn cholla stem that has painfully attached itself to your pant leg sprouted some time around the Battle of Hastings.

Mormon tea | Photo by Jon Sullivan/Flickr/Creative Commons License

Mormon Tea

This odd little plant, known botanically as *Ephedra nevadensis,* was the other "lived too long to measure its age" species in Martin Cody's study referenced just above. This seems fitting: Mormon tea is what people sometimes misleadingly refer to as a "living fossil" in that the vast majority of its close relatives have gone extinct. (One of its cousins that's still around, Welwitschia, is possibly the oddest long-lived desert plant ever, but it isn't native to the Mojave.) Ephedra is a gymnosperm, more closely related to pines and spruces than it is to true flowering plants. It's easily recognized by its oddly jointed, leafless stems bearing either small cones or the scars from former cones at the joints.

Mormon tea is so-called because it has been used as a beverage for both medicinal and recreational purposes: it contains moderate amounts of ephedrine and pseudoephedrine, enough for a mild stimulant (and decongestant) effect. (Don't try this yourself unless you are certain you've got the right plant. Some plants that should not be consumed resemble Ephedra enough to confuse people not well-versed in plant identification. And never take cuttings of any plant on protected or private land without permission.) Mormon tea is an important wildlife food source, with large animals browsing on the stems and smaller ones gathering its seeds. It isn't showy or prominent, unless you're looking for it: it just plugs along, feeding wild things and growing back after it's browsed.

Mormon tea provides a telling indication of just how little we know about even the most common desert plants. While Cody's study indicated that the species may have a very long lifespan, the US Forest Service describes the species' lifespan in frustratingly vague terms as "more than 100 years, many other reputable-seeming sources describe the plant as short-lived -- possibly describing its longevity in cultivation or under heavy grazing pressure. We know so little about this plant, and it's not exactly rare. The desert is truly *terra incognita*, and developing desert wildlands may destroy treasures we don't even know exist.

Turpentine broom's tell-tale flowers | Photo by Joe Decruyenaere/Flickr/Creative Commons License

Turpentine Broom

At first glance, this species -- *Thamnosma montana* -- could be mistaken for Mormon Tea by a beginner: its stem structure is roughly similar, and its leaves are few, small, and temporary. This underscores the importance of getting your plant IDs right before making tea: Turpentine broom has variously been used as an emetic, a laxative, a hallucinogen, and a pesticide.

Fortunately, it isn't really that hard to tell the two species apart. *Thamnosma montana* is a distinctive chartreuse color, and holds small, deep purple flowers up and down its stems. A member of the same plant family (Rutaceae) as citrus, it grows at middle elevations below about 5,500 feet. It's unpalatable to most livestock, though bighorn sheep do eat it. One of turpentine broom's chief ecological values is as a soil-builder: its dense crown of thin stems catches wind-blown organic matter and holds it.

The species is also an important host plant for butterflies. A few years ago on Cima Dome in the Mojave National Preserve, I watched as an unusual hatch of fall butterflies swarmed the desert. One of the species that showed up, the Indra swallowtail, was especially drawn to the abundant turpentine broom there on the Dome, and I watched as the females laid one tiny, jade-colored egg after another on the chartreuse stems, there to hatch out as caterpillars to eat the plant's meager leaves.
One other difference between turpentine broom and ephedra: nobody refers to turpentine broom as "short-lived." Cody's study put an approximate ceiling on the species' longevity in the Mojave Preserve, saying that about five percent of the

individual turpentine broom plants under study would likely live around 1,150 years.

Blackbrush in the Coso Mountains | Photo courtesy BasinandRangeWatch.org

Blackbrush

Alert readers will recall that *Coleogyne ramossissima*, a.k.a. the almost ubiquitous blackbrush, has already been featured here at KCET.org as part of an incredibly long-lived vegetative community. As that previous article explains, a solid cover of blackbrush, which you can find throughout the Mojave at elevations between 2,000 and 5,000 feet, may take as long as 15,000 years to develop, or even longer. Individual blackbrush plants are no slouches in the longevity department, either. Cody's study in the Granite Mountains put blackbrush's top five percent longevity at around 1,250 years.

The list of potentially ancient plants in the Mojave goes on: these are merely some of the most common and easily seen species. You can see hundreds of individuals of every single one of these species on a drive from L.A. to Las Vegas without leaving the Interstate. Many of them are centuries old, or even millennia. All of them are worth learning more about, cherishing and protecting.

Appendix D: Aging Gene Study

The Age-1 and Daf-2 Genes Function in a Common Pathway to Control the Lifespan of Caenorhabditis Elegans

J. R. Dorman, B. Albinder, T. Shroyer, and C. Kenyon

Abstract

Recessive mutations in two genes, daf-2 and age-1, extend the lifespan of Caenorhabditis elegans significantly. The daf-2 gene also regulates formation of an alternative developmental state called the dauer. Here we asked whether these two genes function in the same or different lifespan pathways. We found that the longevity of both age-1 and daf-2 mutants requires the activities of the same two genes, daf-16 and daf-18. In addition, the daf-2(e1370); age-1(hx546) double mutant did not live significantly longer than the daf-2 single mutant. We also found that, like daf-2 mutations, the age-1(hx546) mutation affects certain aspects of dauer formation. These findings suggest that age-1 and daf-2 mutations do act in the same lifespan pathway and extend lifespan by triggering similar if not identical processes.

Aging Cell. 2008 Jan;7(1):13-22. Epub 2007 Nov 7.

Remarkable longevity and stress resistance of nematode PI3K-null mutants.

Ayyadevara S[1], Alla R, Thaden JJ, Shmookler Reis RJ.

Author information
Abstract

The great majority of lifespan-augmenting mutations were discovered in the nematode Caenorhabditis elegans. In particular, genetic disruption of insulin-like signaling extends longevity 1.5- to 3-fold in the nematode, and to lesser degrees in other taxa, including fruit flies and mice. C. elegans strains bearing homozygous nonsense mutations in the age-1 gene, which encodes the class-I phosphatidylinositol 3-kinase catalytic subunit (PI3K(CS)), produce progeny that were thought to undergo obligatory developmental arrest. We now find that, after prolonged developmental times at 15-20 degrees C, they mature into extremely long-lived adults with near-normal feeding rates and motility. They survive to a median of 145-190 days at 20 degrees C, with nearly 10-fold extension of both median and maximum adult lifespan relative to N2DRM, a long-lived wild-type stock into which the null mutant was outcrossed. PI3K-null adults, although a little less thermotolerant, are considerably more resistant to oxidative and electrophilic stresses than worms bearing normal or less long-lived alleles. Their unprecedented factorial gains in survival, under both normal and toxic environments, are attributed to elimination of residual and maternally contributed PI3K(CS) or its products, and consequent modification of kinase signaling cascades.

Appendix E: All 11 Abstracts Mentioning Species Level Selection in the Entire Pub Med Science Database as of 12/2015.

I include these 12 abstracts here to show how little thought has been given to the idea of species selection to date.

Proc Natl Acad Sci U S A. 1975 Feb;72(2):646-50.

A theory of evolution above the species level.

Stanley SM.

Abstract

Gradual evolutionary change by natural selection operates so slowly within established species that it cannot account for the major features of evolution. Evolutionary change tends to be concentrated within speciation events. The direction of transpecific evolution is determined by the process of species selection, which is analogous to natural selection but acts upon species within higher taxa rather than upon individuals within populations. Species selection operates on variation provided by the largely random process of speciation and favors species that speciate at high rates or survive for long periods and therefore tend to leave many daughter species. Rates of speciation can be estimated for living taxa by means of the equation for exponential increase, and are clearly higher for mammals than for bivalve mollusks.

Nature. 1988 May 19;333(6170):214-5.

Species selection and the role of the individual.

Stanley SM.

Proc Natl Acad Sci U S A. 1993 Jan 15;90(2):595-9.

Species selection on variability.

Lloyd EA[1], Gould SJ.

Author information
Abstract

Most analyses of species selection require emergent, as opposed to aggregate, characters at the species level. This "emergent character" approach tends to focus on the search for adaptations at the species level. Such an approach seems to banish the most potent evolutionary property of populations--variability itself--from arguments about species selection (for variation is an aggregate character). We wish, instead, to extend the legitimate domain of species selection to aggregate characters. This extension of selection theory to the species level will concentrate, instead, on the relation between fitness and the species character, whether aggregate or emergent. Examination of the role of genetic variability in the long-term evolution of clades illustrates the cogency of broadening the definition of species selection to include aggregate characters. We reinterpret, in this light, a classic case presented in support of species selection. As originally presented, the species selection explanation of volutid neogastropod evolution was vulnerable to a counter interpretation at the organism level. Once this case is recast within a definition of species selection that reflects the essential structure and broad applicability of hierarchical selection models, the organism-level reinterpretation of variability loses its force. We conclude that species selection on variability is a major force of macroevolution.

J Theor Biol. 1994 Dec 21;171(4):427-30.

Species selection on organismal integration.

Björklund M[1].

Author information
Abstract

Selection processes of entities higher than individuals have recently been suggested to play a potential role in macroevolution. In particular, population level traits such as variability seem likely candidates for higher-level selection processes because they interact with population fitness (survival). In this paper, I expand on that theme and argue that a population level trait, such as organismal integration, measured by the genetic variance-covariance matrix, can be subject to interpopulational selection. This is so because a population consisting of individuals with a high degree of integration will respond faster to selection than a less integrated one, and thus more rapidly reach new optima. This idea generates a number of predictions which are supported by data from natural and laboratory populations of a diverse array of organisms. First, the level of genetic integration in morphological characters is generally quite high. Second, there are a number of cases where the within- and among-population correlation matrices are similarly oriented. Third, the main pattern of morphological variation in birds is that species within genera are most exclusively oriented along a size-axis. These results are consistent with the ideas put forward in this paper, though not providing conclusive evidence.

A genetical theory of species selection.

Rice SH[1].

Author information
Abstract

Species selection, differential rates of speciation or extinction resulting from species level characters, is often invoked as the main mechanism of macroevolution that is not simply an extension of microevolutionary processes. So long as we are careful in defining "species", the logic of species selection is sound. This does not mean, however, that this process can influence evolutionary dynamics under realistic conditions. The principal challenge to the efficacy of species selection as an evolutionary mechanism is the idea that selection between individuals within species will be so much more efficient as to swamp out any effects of selection between species. To assess this, a genetic model is constructed that includes simultaneous selection within and between species, and this is used to ask: under what conditions could species selection influence evolutionary dynamics, even in the face of opposing selection between individuals? The model shows that the efficacy of species selection is strongly determined by the time between speciation events (measured in individual generations), the mutation rate of the character under consideration, and the initial size of a newly formed reproductively isolated population. Data indicate that a few studied lineages have shown sufficiently high speciation rates to make species selection an important mechanism in the evolution of characters with mutation rates on the order of $10^{(-6)}$ per generation. Quantitative characters, such as body size, generally change too readily for species selection to be relevant to their evolution. Complex characters, however, may be good candidates to be influenced by species selection. The interaction of selection within and between species can be subtle, with individual selection looking, from the standpoint of a species, very much like development of an individual. Furthermore, selection between individuals may be the main process assembling complex adaptations, while species selection allows them to persist over long periods of time.

Proc Natl Acad Sci U S A. 1999 Aug 31;96(18):10272-7.

Developmental shifts and species selection in gastropods.

Duda TF Jr[1], Palumbi SR.

Author information
Abstract

The fossil record of marine gastropods has been used as evidence to support the operation of species selection; namely, that species with limited dispersal differentially increase in numbers because they are more likely to speciate than widely dispersing species. This conclusion is based on a tacit phylogenetic assumption that increases in species with limited dispersal are solely the result of speciation within monophyletic groups with low dispersal. To test this assumption, we reconstructed a phylogeny from nuclear sequence data for 70 species of the marine gastropod genus Conus and used it to map the evolution of developmental mode. All eight species without planktonic life history phases recently and independently evolved this characteristic from ancestors with planktonic larval phases, showing that transitions in developmental mode are common in this group. A simple model of species diversification shows that such shifts can control the relative numbers of species with and without dispersing larval stages, leading to apparent species selection. Such results challenge the conclusion that increases in the number of nonplanktonic species relative to species with planktonic larvae over geologic time is necessarily a result of higher rates of speciation of nonplanktonic lineages and show that demonstration of species selection requires a phylogenetic framework.

Evolution. 2003 Feb;57(2):421-7.

Between- and within-host species selection on cytoplasmic incompatibility-inducing Wolbachia in haplodiploids.

Vavre F[1], Fouillet P, Fleury F.

Author information
Abstract

The most common effect of the endosymbiont Wolbachia is cytoplasmic incompatibility (CI), a form of postzygotic reproductive isolation that occurs in crosses where the male is infected by at least one Wolbachia strain that the female lacks. We revisited two puzzling features of Wolbachia biology: how Wolbachia can invade a new species and spread among populations, and how the association, once established in a host species, can evolve, with emphasis on the possible process of infection loss. These questions are particularly relevant in haplodiploid species, where males develop from unfertilized eggs, and females from fertilized eggs. When CI occurs in such species, fertilized eggs either die (female mortality type: FM), or develop into males (male development type: MD), raising one more question: how transition among CI types is possible. We reached the following conclusions: (1) the FM type is a better invader and should be retained preferentially after a new host is captured; (2) given the assumptions of the models, FM and MD types are selected on neither the bacterial side nor the host side; (3) selective pressures acting on both partners are more or less congruent in the FM type, but divergent in the MD type; (4) host and symbiont evolution can drive infection to extinction for all CI types, but the MD type is more susceptible to the phenomenon; and (5) under realistic conditions, transition from MD to FM type is possible. Finally, all these results suggest that the FM type should be more frequent than the MD type, which is consistent with the results obtained so far in haplodiploids.

Theory Biosci. 2010 Sep;129(2-3):113-23. doi: 10.1007/s12064-010-0088-6. Epub 2010 May 26.

Punctuated equilibrium and species selection: what does it mean for one theory to suggest another?

Turner D[1].

Author information
Abstract

McMullin (In: Cohen et al. (eds.) Essays in memory of Imre Lakatos, 1976, In: Leplin (ed.) Scientific realism, 1984) argues that fertility is a theoretical virtue. He thinks of a fertile theory as one whose central metaphors suggest new directions for theoretical development, where those new developments help solve previous problems and anomalies. Nolan (Br J Philos Sci

50:265-282, 1999) argues that fertility in this sense is not a distinctive theoretical virtue in its own right. Rather, Nolan thinks that fertility is reducible to predictive novelty. This article explores the relationship between punctuated equilibrium (PE) and species selection in the light of this philosophical debate about the nature of theoretical fertility, or suggestiveness. I argue that (1) PE suggests, but does not imply, that species selection is a mechanism of evolution; (2) the suggestiveness in this case is not reducible to predictive novelty; (3) species selection is not a metaphorical extension of PE; and (4) getting clear about the way in which PE suggests species selection can help solve a puzzle about punctuated equilibrium. The puzzle is that Eldredge and Gould's initial presentation of PE seems to presuppose a minimalist or extrapolationist view of macroevolution, even though many scientists take PE to challenge that minimalist view.

Syst Biol. 2011 Jul;60(4):503-18. doi: 10.1093/sysbio/syr020. Epub 2011 Apr 2.

Correlations of life-history and distributional-range variation with salamander diversification rates: evidence for species selection.

Eastman JM[1], Storfer A.

Author information
Abstract

Evolutionary biologists have long debated the relative influence of species selection on evolutionary patterns. As a test, we apply a statistical phylogenetic approach to evaluate the influence of traits related to species distribution and life-history characteristics on patterns of diversification in salamanders. We use independent contrasts to test trait-mediated diversification while accommodating phylogenetic uncertainty in relationships among all salamander families. Using a neontological data set, we find several species-level traits to be variable, heritable, and associated with differential success (i.e., higher diversification rates) at higher taxonomic categories. Specifically, the macroecological trait of small geographic-range size is strongly correlated with a higher rate of net diversification. We further consider the role that plasticity in life-history traits appears to fulfill in macroevolutionary processes of lineage divergence and durability. We find that pedotypy--wherein some, but not all, organisms of a species mature in the gilled form without metamorphosing-is also associated with higher net diversification rate than is the absence of developmental plasticity. Often dismissed as an insignificant process in evolution, we provide direct evidence for the role of species selection in lineage diversification of salamanders.

Evolution. 2013 Jun;67(6):1607-21. doi: 10.1111/evo.12083. Epub 2013 Apr 9.

Species selection and the macroevolution of coral coloniality and photosymbiosis.

Simpson C[1].

Author information
Abstract

Differences in the relative diversification rates of species with variant traits are known as species selection. Species selection can produce a macroevolutionary change in the frequencies of traits by changing the relative number of species possessing each trait over time. But species selection is not the only process that can change the frequencies of traits, phyletic microevolution of traits within species and phylogenetic trait evolution among species, the tempo and mode of microevolution can also change trait frequencies. Species selection, phylogenetic, and phyletic processes can all contribute to large-scale trends, reinforcing or canceling each other out. Even more complex interactions among macroevolutionary processes are possible when multiple covarying traits are involved. Here I present a multilevel macroevolutionary framework that is useful for understanding how macroevolutionary processes interact. It is useful for empirical studies using fossils, molecular phylogenies, or both. I illustrate the framework with the macroevolution of coloniality and photosymbiosis in scleractinian corals using a time-calibrated molecular phylogeny. I find that standing phylogenetic variation in coloniality and photosymbiosis deflects the direction of macroevolution from the vector of species selection. Variation in these traits constrains species selection and results in a 200 million year macroevolutionary equilibrium.

Syst Biol. 2015 Nov;64(6):983-99. doi: 10.1093/sysbio/syv046. Epub 2015 Jul 10.

Species Selection Favors Dispersive Life Histories in Sea Slugs, but Higher Per-Offspring Investment Drives Shifts to Short-Lived Larvae.

Krug PJ[1], Vendetti JE[2], Ellingson RA[2], Trowbridge CD[3], Hirano YM[4], Trathen DY[2], Rodriguez AK[2], Swennen C[5],Wilson NG[6], Valdés ÁA[7].

Author information
Abstract

For 40 years, paleontological studies of marine gastropods have suggested that species selection favors lineages with short-lived (lecithotrophic) larvae, which are less dispersive than long-lived (planktotrophic) larvae. Although lecithotrophs appeared to speciate more often and accumulate over time in some groups, lecithotrophy also increased extinction rates, and tests for state-dependent diversification were never performed. Molecular phylogenies of diverse groups instead suggested lecithotrophs accumulate without diversifying due to frequent, unidirectional character change. Although lecithotrophy has repeatedly originated in most phyla, no adult trait has been correlated with shifts in larval type. Thus, both the evolutionary origins of lecithotrophy and its consequences for patterns of species richness remain poorly understood. Here, we test hypothesized links between development mode and evolutionary rates using likelihood-based methods and a phylogeny of 202 species of gastropod molluscs in Sacoglossa, a clade of herbivorous sea slugs. Evolutionary quantitative genetics modeling and stochastic character mapping supported 27 origins of lecithotrophy. Tests for correlated evolution revealed lecithotrophy evolved more often in lineages investing in extra-embryonic yolk, the first adult trait associated with shifts in development mode across a group. However, contrary to predictions from paleontological studies, species selection actually favored planktotrophy; most extant lecithotrophs originated through recent character change, and did not subsequently diversify. Increased offspring provisioning in planktotrophs thus favored shifts to short-lived larvae, which led to short-lived lineages over macroevolutionary time scales. These findings challenge long-standing assumptions about the effects of alternative life histories in the sea. Species selection can explain the long-term persistence of planktotrophy, the ancestral state in most clades, despite frequent transitions to lecithotrophy.

Proc Biol Sci. 2015 Aug 7;282(1812):20151097. doi: 10.1098/rspb.2015.1097.

The extended Price equation quantifies species selection on mammalian body size across the Palaeocene/Eocene Thermal Maximum.

Rankin BD[1], Fox JW[2], Barrón-Ortiz CR[2], Chew AE[3], Holroyd PA[4], Ludtke JA[2], Yang X[2], Theodor JM[2].

Author information
Abstract

Species selection, covariation of species' traits with their net diversification rates, is an important component of macroevolution. Most studies have relied on indirect evidence for its operation and have not quantified its strength relative to other macroevolutionary forces. We use an extension of the Price equation to quantify the mechanisms of body size macroevolution in mammals from the latest Palaeocene and earliest Eocene of the Bighorn and Clarks Fork Basins of Wyoming. Dwarfing of mammalian taxa across the Palaeocene/Eocene Thermal Maximum (PETM), an intense, brief warming event that occurred at approximately 56 Ma, has been suggested to reflect anagenetic change and the immigration of small bodied-mammals, but might also be attributable to species selection. Using previously reconstructed ancestor-descendant relationships, we partitioned change in mean mammalian body size into three distinct mechanisms: species selection operating on resident mammals, anagenetic change within resident mammalian lineages and change due to immigrants. The remarkable decrease in mean body size across the warming event occurred through anagenetic change and immigration. Species selection also was strong across the PETM but, intriguingly, favoured larger-bodied species, implying some unknown mechanism(s) by which warming events affect macroevolution.

A Little More on Species Selection:

So you see Stephen Jay Gould almost gets it….species selection" but not quite in the way I am using it..

It remains controversial among biologists whether selection can operate at and above the level of species. One particular defender of the idea of species selection was Stephen Jay Gould who proposed the view that there exist macroevolutionary processes which shape evolution that are not driven by the microevolutionary mechanisms that are the basis of the [Modern Synthesis](#).[6] If one views species as entities that replicate (speciate) and die (go extinct), then species could be subject to selection and thus could change their occurrence over geological time, much as heritable selected-for traits change theirs over generations.

For evolution to be driven by species selection, differential success must be the result of selection upon species-intrinsic properties, rather than for properties of genes, cells, individuals, or populations within species. Such properties include, for example, population structure, their propensity to speciate, extinction rates, and geological persistence. While the fossil record shows differential persistence of species, examples of species-intrinsic properties subject to natural selection have been much harder to document.

Species selection is the process responsible for the proliferation of species that have lower extinction and higher speciation rates.

Species selection is a reason why macroevolution and microevolution may be uncoupled. Within a species, natural selection favors one character in one species and another in another species; but species selection may cause the species with one of the characters to proliferate, because of the character's consequences for speciation or extinction rates.

This does not mean that the long-term process contradicts, or is incompatible with, the short-term process, only that we cannot understand the long-term evolutionary pattern by studying natural selection in the short-term alone, and extrapolating it. Species selection should not be confused with group selection

Appendix F: It Turns out Werner's Protein (WRN) Shuts Off Aging Genes Just Like All the Other Anti-Aging Proteins.

The following are recent studies of WRN protein about functions I expected to exist-suppressing aging genes, WRN is also involved in differentiation!
AKA aging gene silencing!! It is not just a helicase.
Just like the lamin a protein WRN has a 2nd function in differentiation! viola!

Aging Cell. 2010 Aug;9(4):580-91. doi: 10.1111/j.1474-9726.2010.00585.x. Epub 2010 May 10.

A role for the Werner syndrome protein in epigenetic inactivation of the pluripotency factor Oct4.

Smith JA[1], Ndoye AM, Geary K, Lisanti MP, Igoucheva O, Daniel R.
Author information
Abstract
Werner syndrome (WS) is an autosomal recessive disorder, the hallmarks of which are premature aging and early onset of neoplastic diseases (Orren, 2006; Bohr, 2008). The gene, whose mutation underlies the WS phenotype, is called WRN. The protein encoded by the WRN gene, WRNp, has DNA helicase activity (Gray et al., 1997; Orren, 2006; Bohr, 2008; Opresko, 2008). Extensive evidence suggests that WRNp plays a role in DNA replication and DNA repair (Chen et al., 2003; Hickson, 2003; Orren, 2006; Turaga et al., 2007; Bohr, 2008). However, WRNp function is not yet fully understood. In this study, we show that WRNp is involved in de novo DNA methylation of the promoter of the Oct4 gene, which encodes a crucial stem cell transcription factor. We demonstrate that WRNp localizes to the Oct4 promoter during retinoic acid-induced differentiation of human pluripotent cells and associates with the de novo methyltransferase Dnmt3b in the chromatin of differentiating pluripotent cells. Depletion of WRNp does not affect demethylation of lysine 4 of the histone H3 at the

Oct4 promoter, nor methylation of lysine 9 of H3, but it blocks the recruitment of Dnmt3b to the promoter and results in the reduced methylation of CpG sites within the Oct4 promoter. The lack of DNA methylation was associated with continued, albeit greatly reduced, Oct4 expression in WRN-deficient, retinoic acid-treated cells, which resulted in attenuated differentiation. The presented results reveal a novel function of WRNp and demonstrate that WRNp controls a key step in pluripotent stem cell differentiation.

Horm Res Paediatr. 2010;74(1):33-40. doi: 10.1159/000313366. Epub 2010 Apr 15.

Estrogen prevents senescence through induction of WRN, Werner syndrome protein.

Lee SJ[1], Lee SH, Ha NC, Park BJ.
Author information
Abstract

Werner syndrome is a well-known human progeria. It has been revealed that loss of human WRN is a causal factor of this disease. Since pathological features of Werner syndrome resemble those of menopausal women and become apparent during puberty, we examined the effect of estrogen on WRN gene expression. Here, we reveal that WRN is induced by estrogen but not testosterone. Treatment with estrogen can induce WRN expression at the transcription and translation level in a human breast cell line. Forced expression of the estrogen receptor can restore the responsiveness of WRN to estrogen in a non-responsive cell line. Treatment with estrogen can block DNA damage-induced senescence. Moreover, WRN is suppressed by ATR that is activated by DNA damage, whereas WRN can be induced by ATR elimination. Our results suggest that WRN is essential for prevention of senescence. In addition, our results imply that the reduction of WRN in menopause could be an important factor for menopausal syndrome.

Copyright 2010 S. Karger AG, Basel.

Cell Cycle. 2009 Jul 1;8(13):2080-92. Epub 2009 Jul 5.

The Werner syndrome protein affects the expression of genes involved in adipogenesis and inflammation in addition to cell cycle and DNA damage responses.

Turaga RV[1], Paquet ER, Sild M, Vignard J, Garand C, Johnson FB, Masson JY, Lebel M.
Author information
Abstract
Werner syndrome (WS) is characterized by the premature onset of several age-associated pathologies. The protein deficient in WS (WRN) is a RecQ-type DNA helicase involved in DNA repair, replication, telomere maintenance and transcription. However, precisely how WRN deficiency leads to the numerous WS pathologies is still unknown. Here we use short-term siRNA-based inhibition of WRN to test the direct consequences of its loss on gene expression. Importantly, this short-term knock down of WRN protein level was sufficient to trigger an expression profile resembling fibroblasts established from old donor patients. In addition, this treatment altered sets of genes involved in 14 distinct biological pathways. Besides the already known impact of WRN on DNA replication, DNA repair, the p21/p53 pathway, and cell cycle, gene set enrichment analyses of our microarray data also uncover significant impact on the MYC, E2F, cellular E2A and ETV5 transcription factor pathways as well as adipocyte differentiation, HIF1, NFkappaB and IL-6 pathways. Finally, short-term siRNA-based inhibition of mouse Wrn expression in the pre-adipocyte cell line 3T3-L1 confirmed the impact of WRN on adipogenesis. These results are consistent with the pro-inflammatory status and lipid abnormalities observed in WS patients. This approach thus identified new effectors of WRN activity that might contribute to the WS phenotype.

\

Heterochromatin disorganization associated with premature ageing

Bryony Jones

Nature Reviews Genetics

18 May 2015

Werner syndrome (also known as adult progeria) is a premature ageing disorder with phenotypes such as grey hair, osteoporosis and diabetes, which are linked to defects in mesodermal tissue. Werner syndrome is caused by mutations in the *WRN* gene, which is involved in several fundamental cellular mechanisms, including DNA replication...

"progressive disorganization of heterochromatin underlies the pathology of premature cellular ageing"

Appendix G: The Conservative US government's NIH is Now Promoting the LH/Alzheimer's Disease Connection!

Gonadotropin-releasing hormone receptor system: modulatory role in aging and neurodegeneration. Wang L, Chadwick W, Park SS, Zhou Y, Silver N, Martin B, Maudsley S. *CNS Neurol Disord Drug Targets.* 2010 Nov;9(5):651-60

Receptor Pharmacology Unit, National Institute on Aging, National Institutes of Health, Biomedical Research Center, Baltimore MD 21224, USA.

Abstract

Receptors for hormones of the hypothalamic-pituitary-gonadal axis are expressed throughout the brain. Age-related decline in gonadal reproductive hormones cause imbalances of this axis and many hormones in this axis have been functionally linked to neurodegenerative pathophysiology. Gonadotropin-releasing hormone (GnRH) plays a vital role in both central and peripheral reproductive regulation. GnRH has historically been known as a pituitary hormone; however, in the past few years, interest has been raised in GnRH actions at non-pituitary peripheral targets. GnRH ligands and receptors are found throughout the brain where they may act to control multiple higher functions such as learning and memory function and feeding behavior. The actions of GnRH in mammals are mediated by the activation of a unique rhodopsin-like G protein-coupled receptor that does not possess a cytoplasmic carboxyl terminal sequence. Activation of this receptor appears to mediate a wide variety of signaling mechanisms that show diversity in different tissues. Epidemiological support for a role of GnRH in central functions is evidenced by a reduction in neurodegenerative disease after GnRH agonist therapy. It has previously been considered that these effects were not via direct GnRH action in the brain, however recent data has pointed to a direct central action of these ligands outside the pituitary. We have therefore summarized the evidence supporting a central direct role of GnRH ligands and receptors in controlling central nervous physiology and pathophysiology.

NIH News: New paper suggests elevated LH behind AD

This is a very well referenced and comprehensive review of the literature and data surrounding the concepts of elevated GNRH/LH contributing to AD. Probably most important, it was conducted and prepared by one of the leading neuroscientists at the NIH. Completely independent and with no ties to any private company. Gonadotropin-releasing hormone receptor system: modulatory role in aging and

Reply 1: NIH News!!! New paper suggests elevated LH behind AD

onward replied

"These findings support the premise that GnRH receptor-based therapeutics could be a potential therapeutic target for the treatment of AD. Several double-blind placebo controlled phase II clinical trials are currently underway to conclusively make this determination." Very interesting and encouraging. Thanks for posting, Prodiver. Can anyone find out exactly what "GnRH receptor-based therapeutics...

Reply 2: NIH News!!! New paper suggests elevated LH behind AD

Billstrailrunning replied

These findings support the premise that GnRH receptor-based therapeutics could be a potential therapeutic target for the treatment of AD. Several double-blind placebo controlled phase II clinical trials are currently underway This sounds promising. We will look forward to the results of the intervention. Not sure what the

intervention will be and will it be the same or different for males and
...

Reply 3: NIH News!!! New paper suggests elevated LH behind AD

Prodiver replied

Leuprolide acetate is the compound under study in the Phase II B trials. It is formulated in a patented biopolymer implant, developed by DURECT Corporation. According to the company, it uniquely releases a proprietary dosage level which is much higher than is used in previous applications of the compound to treat prostate cancer, endometriosis or precocious puberty. LA has been shown to be very ...

Reply 4: NIH News!!! New paper suggests elevated LH behind AD

Billstrailrunning replied

Hey ProDiver, great research on your part. I have to say though I'm not thrilled at giving my ADLO Lupron. It is heavy on side-effects. Here is a link: http://www.drugs.com/sfx/leuprolide-side-effects.html. That said, if there is even a hint that it really works I definitely would consider it for my ADLO. Male patients prescribed this medicine are fighting prostate cancer and those I have met are...

Made in the USA
Monee, IL
07 August 2020